SpringerBriefs in Mathematics

SpringerBriefs present concise summaries of cutting-edge research and practical applications across a wide spectrum of fields. Featuring compact volumes of 50 to 125 pages, the series covers a range of content from professional to academic. Briefs are characterized by fast, global electronic dissemination, standard publishing contracts, standardized manuscript preparation and formatting guidelines, and expedited production schedules.

Typical topics might include:

- A timely report of state-of-the art techniques
- A bridge between new research results, as published in journal articles, and a contextual literature review
- A snapshot of a hot or emerging topic
- An in-depth case study
- A presentation of core concepts that students must understand in order to make independent contributions.

SpringerBriefs in Mathematics showcases expositions in all areas of mathematics and applied mathematics. Manuscripts presenting new results or a single new result in a classical field, new field, or an emerging topic, applications, or bridges between new results and already published works, are encouraged. The series is intended for mathematicians and applied mathematicians. All works are peer-reviewed to meet the highest standards of scientific literature.

Titles from this series are indexed by Scopus, Web of Science, Mathematical Reviews, and zbMATH.

Marek Galewski · Dumitru Motreanu

Competing Operators and Their Applications to Boundary Value Problems

Marek Galewski [iD]
Institute of Mathematics
Łódź University of Technology
Łódź, Poland

Dumitru Motreanu
Département de Mathématiques
Universite de Perpignan
Perpignan, France

ISSN 2191-8198   ISSN 2191-8201   (electronic)
SpringerBriefs in Mathematics
ISBN 978-3-032-15444-6   ISBN 978-3-032-15445-3   (eBook)
https://doi.org/10.1007/978-3-032-15445-3

This Springer imprint is published by the registered company Springer Nature Switzerland AG
The registered company address is: Gewerbestrasse 11, 6330 Cham, Switzerland

If disposing of this product, please recycle the paper.

Preface

The purpose of this text is to present certain techniques that have recently emerged in the study of nonlinear operator equations. These methods address the solvability of nonlinear problems that fall outside the scope of classical monotonicity frameworks. Specifically, we consider equations that neither satisfy monotonicity nor pseudomonotonicity conditions, especially in the non-potential setting where the governing operator cannot be derived from a variational principle. The associated energy functionals, if they exist, often fail to possess sequential weak lower semicontinuity. Such methods are motivated by a wide range of applications in nonlinear partial differential equations, variational inequalities, and differential inclusions, where classical methods prove insufficient.

Łódź, Poland   Marek Galewski
Perpignan, France   Dumitru Motreanu
May 2025

Introduction

In this text, we are concerned with nonlinear problems driven by differential operators that do not satisfy neither monotonicity nor pseudomonotonicity assumptions. Specifically, given a reflexive Banach space E, we study equations of the form

$$A(u) = F(u),$$

interpreted in the weak sense, where $A, F : E \rightarrow E^*$ are bounded and continuous operators such that the difference $A - F$ is coercive. In typical applications, E is a suitable Sobolev space, A is a second-order nonlinear differential operator that is neither monotone nor pseudomonotone, and F is a Nemytskij-type operator induced by convection, possibly depending on the gradient. This structure renders the problem non-potential or, alternatively, variational but lacking standard compactness properties.

The analytical methods employed rely fundamentally on coercivity and continuity, allowing for the construction of approximation schemes whose convergence is guaranteed by the coercivity of the operator. The nature of the approximation depends on whether the problem admits a variational formulation. In the variational case, we utilize Ritz-type approximations obtained via minimization of the Euler action functional over finite-dimensional subspaces. In the non-variational case, we construct Galerkin-type approximations. However, due to the absence of monotonicity, convergence of these finite-dimensional schemes to a weak solution is not guaranteed. Nevertheless, coercivity ensures boundedness of the approximating sequence, which in turn implies weak convergence (at least for a subsequence).

This observation motivates the introduction of a new notion of solution, referred to as a *generalized solution*, defined as the weak limit of a sequence generated by finite-dimensional approximation schemes. The existence of such solutions can be established under minimal structural assumptions, provided that the operator driving the equation is at least pseudomonotone, coercive and continuous. To formalize this approach, we introduce an abstract existence result that serves as a counterpart to the Browder–Minty Theorem in the non-variational setting and to the Weierstrass–Tonelli Theorem in the variational case. This framework necessitates the adaptation of several classical techniques to accommodate the lack of monotonicity.

We establish two abstract existence results and apply them to nonlinear boundary value problems. In particular, we consider problems involving unbounded weights, which require the implementation of suitable truncation techniques to ensure well-posedness.

Given the reliance of our analysis on abstract tools from monotone operator theory and variational methods, we include a dedicated chapter that reviews these foundational techniques, complete with proofs and commentary to facilitate a deeper understanding of the generalized solution concept. Preceding this, we provide a chapter on the necessary background in Sobolev spaces.

The main results are presented in Chaps. 3–5. Chapter 3 introduces the notion of generalized solution in the non-variational setting. We begin with the construction of a Galerkin-type approximation scheme for a model problem and proceed to develop an abstract framework for such problems. This reflects the historical development of the theory, which evolved from concrete applications to a more general abstract formulation. Chapter 4 is devoted to the existence theory for variational problems. We begin with an abstract result and then proceed to applications.

Finally, in Chap. 5, we extend our analysis to the case of *differential inclusions*, where the right-hand side of the equation is set-valued. This generalization allows us to treat a broader class of problems, including those with discontinuous or multivalued nonlinearities, and highlights the flexibility of the generalized solution framework in handling nonstandard phenomena.

Contents

Chapter 1
Background from Function Spaces

Abstract In this chapter, we provide some useful information on the function space setting, which we will require in the sequel, involving Sobolev spaces and weighted Sobolev spaces, as well as an overview of variable Sobolev spaces and an introduction to fractional Sobolev spaces. We do not give proofs, apart from some embedding theorems, for any of the results we introduce, as this section is intended to set the background for our further discussions.

Keywords Embeddings · Fractional spaces · Niemytskij operator · Sobolev spaces · Variable Sobolev spaces

1.1 The Composition Operator and Its Continuity

Let Ω be an open subset of $\mathbb{R}^N$ with $N \geq 1$ and let an integer $m \geq 1$.

Definition 1.1 (*Carathéodory function*) We say that $f : \Omega \times \mathbb{R}^m \to \mathbb{R}^m$ is a Carathéodory function if the following conditions are satisfied:

(i) $x \mapsto f(x, u)$ is measurable on Ω for each fixed $u \in \mathbb{R}^m$;
(ii) $u \mapsto f(x, u)$ is continuous on $\mathbb{R}^m$ for almost all $x \in \Omega$.

Denote by $\mathcal{M}$ the set of all measurable functions $u : \Omega \to \mathbb{R}^m$. Generally, the composition of measurable functions need not be measurable, but when $u \in \mathcal{M}$ and $f : \Omega \times \mathbb{R}^m \to \mathbb{R}^m$ is a Carathéodory function, it follows that $x \mapsto f(x, u(x))$ is measurable over Ω. The Niemytskij operator induced by f is the map

$$N_f : \mathcal{M} \to \mathcal{M}$$

given pointwise for $u \in \mathcal{M}$ by

$$N_f(u)(\cdot) = f(\cdot, u(\cdot)) \text{ a.e. on } \Omega.$$

M. Galewski and D. Motreanu, *Competing Operators and Their Applications to Boundary Value Problems*, SpringerBriefs in Mathematics,
https://doi.org/10.1007/978-3-032-15445-3_1

We have the following result regarding the continuity of Niemytskij operators. For related results we refer to [23].

Theorem 1.1 (Krasnosel'skii Theorem) *Let $f : \Omega \times \mathbb{R}^m \to \mathbb{R}^m$ be a Carathéodory function for which there exist constants $b \geq 0$ and p_1, $p_2 \geq 1$, and a nonnegative function $a \in L^{p_2}(\Omega)$ such that*

$$|f(x, u)| \leq a(x) + b|u|^{\frac{p_1}{p_2}} \ for\ a.e.\ x \in \Omega\ and\ all\ u \in \mathbb{R}^m.$$

Then the Niemytskij operator $N_f : L^{p_1}(\Omega, \mathbb{R}^m) \to L^{p_2}(\Omega, \mathbb{R}^m)$ induced by f is well defined (i.e. $N_f(u) \in L^{p_2}(\Omega)$ for all $u \in L^{p_1}(\Omega)$) bounded (in the sense it maps bounded sets into bounded sets) and continuous.

Remark 1.1 In the above considerations, one can assume that the set Ω is a measurable subset of $\mathbb{R}^N$. We assumed that it is open due to the fact that in what follows we work only with open sets with a suitably regular boundary.

1.2 On the (Weighted) Sobolev Spaces and Embeddings

We start by presenting some basic elements of the theory of (weighted) Sobolev spaces that will be used subsequently. More details can be found in [17].

Fix a real number $p \in (1, +\infty)$, a nonnegative function $a \in L^1(\Omega)$ and a bounded domain $\Omega \subset \mathbb{R}^N$ of Lebesgue measure $|\Omega|$ and with Lipschitz boundary $\partial\Omega$. We introduce the weighted Sobolev space $W^{1,p}(a, \Omega)$ (with the weight a) as

$$W^{1,p}(a, \Omega) := \left\{ u \in L^p(\Omega) : \int_\Omega a(x)|\nabla u(x)|^p \mathrm{d}x < \infty \right\}.$$

Here the notation ∇u stands for the gradient of $u \in L^p(\Omega)$ in the sense of distributions. Precisely speaking, let $u \in L^1_{\mathrm{loc}}(\Omega)$ be a locally summable function. If for some $1 \leq i \leq N$ there exists a locally summable function $v \in L^1_{\mathrm{loc}}(\Omega)$ such that

$$\int_\Omega u(x) \frac{\partial \varphi(x)}{\partial x_i} \mathrm{d}x = - \int_\Omega v(x) \varphi(x) \mathrm{d}x \quad \text{for all test functions } \varphi \in C_c^\infty(\Omega)$$

then we say that v is the weak partial derivative of u with respect to variable x_i and write

$$v = \frac{\partial u}{\partial x_i} \text{ or } v = u_{x_i}.$$

If all weak partial derivatives exist we say that the function u is weakly differentiable and define the weak gradient as follows

$$\nabla u = \left[\frac{\partial}{\partial x_1} u, \ldots, \frac{\partial}{\partial x_N} u \right].$$

It is proved that $W^{1,p}(a, \Omega)$ is a Banach space endowed with the norm

$$\|u\|_{W^{1,p}(a,\Omega)} := \left(\|u\|_{L^p(\Omega)}^p + \int_\Omega a(x)|\nabla u(x)|^p dx \right)^{\frac{1}{p}}, \quad \forall u \in W^{1,p}(a, \Omega).$$

If $a(x) \equiv 1$, then $W^{1,p}(a, \Omega)$ reduces to the ordinary Sobolev space $W^{1,p}(\Omega)$. We refer to [8] for a suitable background regarding standard Sobolev spaces (see also [1]).

We note that

$$C_c^\infty(\Omega) \subset W^{1,p}(a, \Omega).$$

The space $W_0^{1,p}(a, \Omega)$ is next defined as the closure of $C_c^\infty(\Omega)$ in $W^{1,p}(a, \Omega)$ with respect to the norm $\| \cdot \|_{W^{1,p}(a,\Omega)}$. Denote by $W_0^{1,p}(a, \Omega)^*$ the dual space of $W_0^{1,p}(a, \Omega)$.

In the case where $a(x) \equiv 1$, $W_0^{1,p}(a, \Omega)$ becomes the Sobolev space $W_0^{1,p}(\Omega)$, whilst the dual $W_0^{1,p}(a, \Omega)^*$ turns to be $W^{-1,p'}(\Omega)$ with

$$p' := \frac{p}{p-1}$$

being the co called conjugate exponent. Set

$$p^* := \frac{Np}{N-p} \text{ if } N > p.$$

called the Sobolev critical exponent. The Sobolev embedding theorem ensures the continuous embedding of $W_0^{1,p}(\Omega)$ into $L^{p^*}(\Omega)$ denoted as

$$W_0^{1,p}(\Omega) \hookrightarrow L^{p^*}(\Omega)$$

and the Rellich-Kondrachov theorem guarantees what follows:

(i) The embedding $W_0^{1,p}(\Omega) \subset L^q(\Omega)$ is compact for all $q \in [1, p^*)$, which we denote by

$$W_0^{1,p}(\Omega) \hookrightarrow\hookrightarrow L^q(\Omega)$$

(ii) If $p = N$, the inclusion $W_0^{1,p}(\Omega) \subset L^q(\Omega)$ is compact for all $q \in [1, +\infty)$.

(iii) If $p > N$, there is the compact embedding $W_0^{1,p}(\Omega) \subset C(\overline{\Omega})$ (Morrey's theorem).

A compact embedding theorem for the weighted space $W_0^{1,p}(a, \Omega)$ can be obtained under the following hypothesis taken from [17, page 26]:

$$\textbf{(H)} \quad a^{-s} \in L^1(\Omega) \ \text{ for some } s \in \left(\max \left\{ \frac{N}{p}, \frac{1}{p-1} \right\}, +\infty \right). \tag{1.1}$$

Theorem 1.2 (Compact embedding for weighted spaces) *Under assumption **(H)**, there is the compact embedding*

$$W_0^{1,p}(a, \Omega) \hookrightarrow\hookrightarrow L^p(\Omega). \tag{1.2}$$

Furthermore, we have that

$$\|u\|_{W_0^{1,p}(a,\Omega)} := \left(\int_\Omega a(x)|\nabla u(x)|^p dx \right)^{\frac{1}{p}}, \quad \forall u \in W_0^{1,p}(a, \Omega),$$

is an equivalent norm on $W_0^{1,p}(a, \Omega)$ for which $W_0^{1,p}(a, \Omega)$ becomes a uniformly convex (and thus reflexive) Banach space.

Proof With an s chosen according to hypothesis (H), see (1.1), we set

$$p_s := \frac{ps}{s+1}. \tag{1.3}$$

Let us note from (1.3) that $p_s > 1$, $p_s < p$, $p_s/(p - p_s) = s$. Hölder's inequality and hypothesis (H) imply

$$\int_\Omega |\nabla u(x)|^{p_s} dx = \int_\Omega \left(a(x)^{\frac{p_s}{p}} |\nabla u(x)|^{p_s} \right) a(x)^{-\frac{p_s}{p}} dx$$

$$\leq \left(\int_\Omega a(x)|\nabla u(x)|^p dx \right)^{\frac{p_s}{p}} \left(\int_\Omega a(x)^{-\frac{p_s}{p-p_s}} dx \right)^{\frac{p-p_s}{p}}$$

$$\leq \left\| a^{-s} \right\|_{L^1(\Omega)}^{\frac{1}{s+1}} \|u\|_{W^{1,p}(a,\Omega)}^{p_s}$$

for all $u \in W^{1,p}(a, \Omega)$. Therefore there is the continuous embedding

$$W_0^{1,p}(a, \Omega) \hookrightarrow W_0^{1,p_s}(\Omega). \tag{1.4}$$

On the other hand, the Rellich-Kondrachov embedding theorem ensures the compact embedding

$$W_0^{1,p_s}(\Omega) \hookrightarrow\hookrightarrow L^r(\Omega),$$

with

$$1 \le r < p_s^* := \frac{Np_s}{N - p_s} \quad if \quad N > p_s$$

and

$$1 \le r < +\infty \ if \ N \le p_s.$$

Condition (H) guarantees that $s > N/p$. Then it is straightforward to see that $p_s^* > p$. Hence there holds the compact embedding

$$W_0^{1,p_s}(\Omega) \hookrightarrow\hookrightarrow L^p(\Omega).$$

Then, in view of (1.4), we have the compact embedding (1.2).

Due to the fact that the embedding (1.2) is continuous, it is clear that the norm $\|\cdot\|_{W_0^{1,p}(a,\Omega)}$ is equivalent to the norm $\|\cdot\|_{W^{1,p}(a,\Omega)}$ on $W_0^{1,p}(a,\Omega)$.

It remains to show that the Banach space $W_0^{1,p}(a,\Omega)$ is uniformly convex. By assumption (H), we know that $a^{-s} \in L^1(\Omega)$ with some $s > 1/(p-1)$. We infer that

$$a^{-\frac{1}{p-1}} \in L^1(\Omega)$$

because of the estimate

$$\int_\Omega a(x)^{-\frac{1}{p-1}}\mathrm{d}x = \int_{\{a(x)<1\}} a(x)^{-\frac{1}{p-1}}\mathrm{d}x + \int_{\{a(x)\ge 1\}} a(x)^{-\frac{1}{p-1}}\mathrm{d}x$$

$$\le \int_\Omega a(x)^{-s}\mathrm{d}x + |\Omega| < \infty.$$

Then from [17, Theorem 1.3] it turns out that $W_0^{1,p}(a,\Omega)$ is uniformly convex, thus completing the proof. $\qquad\square$

Corollary 1.1 (Compact embedding for Sobolev spaces) *There is the compact embedding*

$$W_0^{1,p}(\Omega) \hookrightarrow\hookrightarrow L^p(\Omega).$$

In addition, we have that

$$\|u\|_{W_0^{1,p}(\Omega)} := \left(\int_\Omega |\nabla u(x)|^p dx\right)^{\frac{1}{p}}, \quad \forall u \in W_0^{1,p}(\Omega),$$

is an equivalent norm on $W_0^{1,p}(\Omega)$ for which $W_0^{1,p}(\Omega)$ is a uniformly convex Banach space.

Proof It suffices to apply Theorem 1.2 for $a(x) \equiv 1$. $\square$

In case $p = 2$ we will employ notation

$$H_0^1(\Omega) := W_0^{1,2}(\Omega) \text{ and } H^1(\Omega) := W^{1,2}(\Omega).$$

Now we focus on the embedding of weighted Sobolev spaces involving two nonnegative weights $a \in L^1(\Omega)$ and $b \in L^1(\Omega)$.

Proposition 1.1 (Continuous embedding for weighted spaces) *Let* $1 < q < p < +\infty$ *and* $a, b \in L^1(\Omega)$. *If it holds*

$$a^{-\frac{q}{p-q}} b^{\frac{p}{p-q}} \in L^1(\Omega), \tag{1.5}$$

then there is the continuous embedding

$$W^{1,p}(a, \Omega) \hookrightarrow W^{1,q}(b, \Omega).$$

Proof From assumption (1.5) and Hölder's inequality we derive

$$\int_\Omega b(x)|\nabla u(x)|^q \mathrm{d}x = \int_\Omega (a(x)^{-\frac{q}{p}} b(x))(a(x)^{\frac{q}{p}} |\nabla u(x)|^q) \mathrm{d}x$$

$$\leq \left(\int_\Omega a(x)^{-\frac{q}{p-q}} b(x)^{\frac{p}{p-q}} \mathrm{d}x \right)^{\frac{p-q}{p}} \left(\int_\Omega a(x)|\nabla u(x)|^p \mathrm{d}x \right)^{\frac{q}{p}}$$

$$\leq \left\| a^{-\frac{q}{p-q}} b^{\frac{p}{p-q}} \right\|_{L^1(\Omega)}^{\frac{p-q}{p}} \|u\|_{W^{1,p}(a,\Omega)}^q$$

for all $u \in W^{1,p}(a, \Omega)$. The desired conclusion is achieved. $\square$

For additional results we refer to [17, 18, 29, 72].

1.3 On the Variable Exponent Spaces

Let $\Omega \subseteq \mathbb{R}^N$ with $N \geq 2$ be a bounded domain whose boundary is Lipschitz continuous. Unless stated otherwise, we assume that $p \in C(\bar{\Omega})$ satisfies

$$p(x) > 1 \text{ for all } x \in \bar{\Omega}.$$

In accordance with the classical Weierstrass-Tonelli Theorem, we define:

$$p^- := \min_{x \in \bar{\Omega}} p(x) \quad \text{and} \quad p^+ := \max_{x \in \bar{\Omega}} p(x).$$

We denote by $p' \in C(\bar{\Omega})$ the conjugate variable exponent corresponding to the function p, defined pointwise by:

$$\frac{1}{p(x)} + \frac{1}{p'(x)} = 1 \quad \text{for all } x \in \bar{\Omega}.$$

Let $\mathcal{M}$ denote the set of all measurable functions $u : \Omega \to \mathbb{R}$. The modular associated with the exponent function p is given by:

$$\rho(u) := \int_{\Omega} |u(x)|^{p(x)} \, dx \quad \text{for all } u \in \mathcal{M}.$$

The variable exponent Lebesgue space $L^{p(x)}(\Omega)$ is defined as:

$$L^{p(x)}(\Omega) := \{u \in M(\Omega) : \rho(u) < +\infty\},$$

and is equipped with the Luxemburg norm:

$$\|u\|_{p(x)} := \inf \left\{\lambda > 0 : \rho\left(\frac{u}{\lambda}\right) \leq 1\right\} \quad \text{for all } u \in L^{p(x)}(\Omega).$$

A generalized Hölder-type inequality holds in this setting:

$$\int_{\Omega} |u(x)v(x)| \, dx \leq \left[\frac{1}{p^-} + \frac{1}{(p')^-}\right] \|u\|_{p(x)} \|v\|_{p'(x)} \leq 2\|u\|_{p(x)} \|v\|_{p'(x)},$$

for all $u \in L^{p(x)}(\Omega)$ and $v \in L^{p'(x)}(\Omega)$.

Basic properties of the space $L^{p(x)}(\Omega)$ are summarized below:

Lemma 1.1 (Continuous embedding) *The space $\left(L^{p(x)}(\Omega), \|\cdot\|_{p(x)}\right)$ is a separable, uniformly convex, and hence reflexive Banach space. Its dual space is given by $L^{p'(x)}(\Omega)$. Moreover, for any $p_1, p_2 \in C(\bar{\Omega})$ such that*

$$1 < p_1(x) \leq p_2(x) \text{ for all } x \in \bar{\Omega},$$

we have the continuous embedding:

$$L^{p_2(x)}(\Omega) \hookrightarrow L^{p_1(x)}(\Omega).$$

Remark 1.2 From Lemma 1.1, we immediately obtain the following continuous embeddings:

$$L^{p^+}(\Omega) \hookrightarrow L^{p(x)}(\Omega) \hookrightarrow L^{p^-}(\Omega).$$

Although the modular ρ and the norm $\|\cdot\|_{p(x)}$ are closely related, a key distinction from the constant exponent case is that the modular is no longer a power of the norm.

Proposition 1.2 (Modular vs. norm) *Let $p \in C(\bar{\Omega})$ be such that $p(x) > 1$ for all $x \in \bar{\Omega}$. Then the following statements hold:*

(i) $\|u\|_{p(x)} < 1$ *(resp. > 1, $= 1$) if and only if $\rho(u) < 1$ (resp. > 1, $= 1$),*

(ii) *If $\|u\|_{p(x)} < 1$, then $\|u\|_{p(x)}^{p^+} \leq \rho(u) \leq \|u\|_{p(x)}^{p^-}$,*

(iii) *If $\|u\|_{p(x)} > 1$, then $\|u\|_{p(x)}^{p^-} \leq \rho(u) \leq \|u\|_{p(x)}^{p^+}$.*

Remark 1.3 From Proposition 1.2, we deduce that

$$\|u\|_{p(x)}^{p^-} - 1 \leq \rho(u) \leq \|u\|_{p(x)}^{p^+} + 1$$

holds for all $u \in L^{p(x)}(\Omega)$. Furthermore, we observe that

$$\rho(u) \geq r(u)\|u\|_{p(x)},$$

for all $u \in L^{p(x)}(\Omega)$, where $r : [0, +\infty) \to [0, +\infty)$ is a continuous, increasing function defined by

$$r(t) = \begin{cases} t^{p^+ - 1}, & 0 \leq t \leq 1 \\ t^{p^- - 1}, & t > 1 \end{cases} \tag{1.6}$$

Analogously to the classical Sobolev space $W^{1,p}(\Omega)$ for constant exponents, we define the variable exponent Sobolev space $W^{1,p(x)}(\Omega)$ as

$$W^{1,p(x)}(\Omega) := \left\{ u \in L^{p(x)}(\Omega) : |\nabla u| \in L^{p(x)}(\Omega) \right\}.$$

This space is equipped with the norm

$$\|u\|_{1,p(x)} := \|u\|_{p(x)} + \|\nabla u\|_{p(x)},$$

for all $u \in W^{1,p(x)}(\Omega)$, where $\|\nabla u\|_{p(x)} = \| |\nabla u| \|_{p(x)}$.

As usual, we define $W_0^{1,p(x)}(\Omega)$ as the closure of $C_0^\infty(\Omega)$ with respect to the norm $\|\cdot\|_{1,p(x)}$. By Lemma 1.1, both $W^{1,p(x)}(\Omega)$ and $W_0^{1,p(x)}(\Omega)$ are uniformly convex, separable, and reflexive Banach spaces.

Moreover, the Poincaré inequality holds in $W_0^{1,p(x)}(\Omega)$, meaning that there exists a constant $c_p > 0$ such that:

$$\|u\|_{p(x)} \leq c_p \|\nabla u\|_{p(x)}, \quad \text{for all} \ \ u \in W_0^{1,p(x)}(\Omega).$$

Consequently, we may consider the equivalent norm:

$$\|u\| := \|\nabla u\|_{p(x)} \quad \text{for all } u \in W_0^{1,p(x)}(\Omega).$$

In line with the classical Sobolev embedding theorems, we have the following result:

Proposition 1.3 (Embeddings for variable exponent Sobolev spaces) *Assume that* $N > p^+$. *Let the function* $q \in C(\bar{\Omega})$ *be such that:*

$$1 < q(x) \leq p^*(x) := \frac{Np(x)}{N - p(x)}$$

for all $x \in \bar{\Omega}$. *Then, the following continuous embedding holds:*

$$W_0^{1,p(x)}(\Omega) \hookrightarrow L^{q(x)}(\Omega).$$

Moreover, if

$$1 < q(x) < p^*(x) \text{ for all } x \in \bar{\Omega},$$

then the above embedding is compact.

Note that the function p^* represents the critical Sobolev exponent associated with the variable exponent function p.

For a comprehensive study of variable Sobolev spaces we refer to [14, 64, 70].

1.4 On the Fractional Sobolev Spaces

Let Ω be a bounded domain in $\mathbb{R}^N$, $N \geq 2$, with a Lipschitz boundary $\partial\Omega$. Given $s \in (0, 1)$, the Gagliardo seminorm of a measurable function $u : \mathbb{R}^N \to \mathbb{R}$ is

$$[u]_{s,p} := \left(\int_{\mathbb{R}^N \times \mathbb{R}^N} \frac{|u(x) - u(y)|^p}{|x - y|^{N+ps}} \, dx \, dy \right)^{1/p}.$$

We define the fractional Sobolev space

$$W^{s,p}(\mathbb{R}^N) := \left\{ u \in L^p(\mathbb{R}^N) : [u]_{s,p} < +\infty \right\}$$

and endow it with the norm

$$\|u\|_{W^{s,p}(\mathbb{R}^N)} := \left(\|u\|_{L^p(\mathbb{R}^N)}^p + [u]_{s,p}^p \right)^{1/p}.$$

The space

$$W_0^{s,p}(\Omega) := \left\{ u \in W^{s,p}(\mathbb{R}^N) : u = 0 \text{ a.e. in } \mathbb{R}^N \setminus \Omega \right\}$$

is endowed with the equivalent norm

$$\|u\|_{s,p} := [u]_{s,p}, \quad u \in W_0^{s,p}(\Omega).$$

The dual space of $W_0^{s,p}(\Omega)$ is $W^{-s,p'}(\Omega)$.

Remark 1.4 Note that we abuse notation a bit while defining $W_0^{s,p}(\Omega)$, since all members of this space in fact are defined over $\mathbb{R}^N$.

The fractional Sobolev critical exponent is

$$P_s^* := \frac{Np}{N - sp} \quad if \quad sp < N$$

and $P_s^* = +\infty$ otherwise. From Propositions 2.1–2.2, Theorem 6.7, and Corollary 7.2 in [61] we have the following result.

Proposition 1.4 (Embeddings for fractional Sobolev spaces) *If* $1 \le p < +\infty$ *then the following hold:*

(a) *If* $0 < s' \le s'' < 1$, *then* $W_0^{s'',p}(\Omega) \hookrightarrow W_0^{s',p}(\Omega)$ *continuously.*
(b) *The embedding* $W_0^{s,p}(\Omega) \hookrightarrow L^r(\Omega)$ *is continuous for all* $r \in [1, P_s^*]$ *and compact when* $r < P_s^* < +\infty$.

Remark 1.5 In this respect, we cite from [48] (see also [54]) that, contrary to the non-fractional case,

$$s \in (0, 1), \quad 1 \le q < p \le +\infty \quad \nRightarrow \quad W_0^{s,p}(\Omega) \subseteq W_0^{s,q}(\Omega).$$

A comprehensive study of fractional spaces is to be found in a recent book [20]. We also refer to [49, 61].

Chapter 2
Aresume on Existence Methods

Abstract In this chapter, we introduce some fundamental tools essential for our subsequent abstract approach. We begin by introducing the Weierstrass–Tonelli theorem, which provides a foundation for the variational treatment of nonlinear problems, as well as the Ekeland variational principle. Then we proceed to the theory of monotone operators, supplementing it with numerous comments and examples. We also present the Browder–Minty theorem and the surjectivity theorem for pseudomonotone, bounded, and continuous operators. We place particular emphasis on the solvability of Galerkin-type approximations, which is why the finite-dimensional case is thoroughly covered. We present only those proofs whose arguments are crucial for the later parts of this text and are vital for understanding the concept of generalized solutions and the issues related to their existence. Some examples of standard applications are also given throughout the chapter.

Keywords Browder–Minty theorem · Ekeland variational principle · Fractional p-Laplacian · Gâteaux derivative · Galerkin basis · Monotone operator · d-monotone operator · pseudomonotone operator · PS-sequence · p-Laplacian · $p(x)$-Laplacian · $(S)_+$-property · The Weierstrass–Tonelli theorem

2.1 On the Direct Method in the Calculus of Variation

In the whole chapter, if not said otherwise, E is a real, separable, and reflexive Banach space. By E^* we understand its dual, and $<, >$ stands for the duality pairing.

The variational approach to solving nonlinear equations relies on the minimization of associated action functionals. This method connects the solutions of the equations to the minima of these action functionals. The foundation of this approach is the renowned Weierstrass Theorem (sometimes called the Weierstrass-Tonelli Theorem), which guarantees the existence of a global minimum for a continuous functional over a compact set.

© The Author(s), under exclusive license to Springer Nature Switzerland AG 2026
M. Galewski and D. Motreanu, *Competing Operators and Their Applications to Boundary Value Problems*, SpringerBriefs in Mathematics,
https://doi.org/10.1007/978-3-032-15445-3_2

In more detail, the variational method involves defining an action functional whose critical points correspond to the weak solutions of the nonlinear equations. This approach is particularly powerful in the context of nonlinear problems with convection which does not depend on the gradient.

The Weierstrass Theorem plays a crucial role in this framework by ensuring that a global minimum exists, thus providing a solid foundation for the variational approach. This theorem states that any continuous functional defined on a compact set will achieve its minimum value, which is essential for proving the existence of solutions to the nonlinear equations.

Theorem 2.1 (Weierstrass Theorem) *Let $J : E \to \mathbb{R}$ be a lower semicontinuous functional and let $D \subset E$ be a compact subset of a Banach space E. Then J has at least one argument of a minimum over D.*

The applicability of Weierstrass Theorem in infinite-dimensional setting is restricted by the fact that the compact sets in infinite dimensional spaces have empty interiors. In order to overcome this, we pass to sequentially weakly lower semicontinuous functionals.

Definition 2.1 (*Sequential weak lower semicontinuity*) We say that a functional $J : E \to \mathbb{R}$ is sequentially weakly lower semicontinuous (sometimes called s.w.l.s.c. for short) if for any $x_0 \in E$ and any sequence $\{x_n\}_{n \geq 1} \subset E$ such that $x_n \rightharpoonup x_0$ it holds

$$\liminf_{n \to +\infty} J(x_n) \geq J(x_0).$$

An equivalent condition is that the Lebesgue set

$$J^\alpha := \{x \in E : J(x) \leq \alpha\}$$

be sequentially weakly closed for all $\alpha \in \mathbb{R}$. A sufficient condition for the above to hold involves convexity and lower semicontinuity.

Theorem 2.2 (Sufficient condition for s.w.l.s.c.) *Assume that the functional $J : E \to \mathbb{R}$ is lower semicontinuous and convex. Then J is sequentially weakly lower semicontinuous.*

Proof Since J is lower semicontinuous, it follows that the sets J^α are closed for all $\alpha \in \mathbb{R}$. Since J is convex, it follows that sets J^α are also convex. In view of the Mazur Lemma, we see that all J^α are sequentially weakly closed, so the conclusion is achieved. $\qquad\qquad\square$

Theorem 2.2 ensures that in uniformly convex spaces we have the following property:

Lemma 2.1 *Let E be a uniformly convex space. Let $\{x_n\}_{n \geq 1} \subset E$ and an element $x_0 \in E$. If $x_n \rightharpoonup x_0$ and $\|x_n\| \to \|x_0\|$, then $x_n \to x_0$.*

Remark 2.1 (*On the comparison of weak semicontinuity and norm semicontinuity*) Notice that the weak sequential continuity of a functional is a stronger notion than continuity. For example, a continuous function $J : l^2 \to \mathbb{R}$ defined by

$$J(x) = -\|x\|,$$

is not sequentially weakly lower semicontinuous. If it were such, then $J_1(x) = \|x\|$ would be sequentially weakly upper semicontinuous and hence sequentially weakly continuous. This is not true because for the sequence $\{e_n\}_{n\geq 1}$ of unit vectors weakly convergent to 0_{l^2}, we have that $J_1(e_n) = 1$. Here l^2 is the Hilbert space of numerical series summable with square and endowed with a standard scalar product.

We provide the following counterpart of the Weierstrass Theorem.

Theorem 2.3 (Weierstrass-Tonelli Theorem) *Let $J : E \to \mathbb{R}$ be a sequentially weakly lower semicontinuous functional and let $D \subset E$ be a sequentially weakly compact set. Then J has at least one argument of a minimum over D.*

Proof Suppose that the funtctional J is not bounded from below on D. Then there is a sequence $\{x_n\}_{n\geq 1} \subset D$ such that

$$\lim_{n \to +\infty} J(x_n) = -\infty.$$

Observe that $\{x_n\}_{n\geq 1}$ has a weakly convergent subsequence $\left\{x_{n_k}\right\}_{k=1}^{\infty}$ with a weak limit $\tilde{x}$. Hence

$$\lim_{k \to +\infty} J\left(x_{n_k}\right) = -\infty.$$

Since J is sequentially weakly lower semicontinuous, we see that

$$-\infty = \lim_{k \to +\infty} J\left(x_{n_k}\right) \geq J(\tilde{x}),$$

which is impossible. Hence J is bounded from below on D and it has a weakly convergent minimizing sequence $\{x_n\}_{n\geq 1} \subset D$ with a weak limit x_0. We then have

$$\inf_{x \in D} J(x) = \liminf_{n \to +\infty} J(x_n) \geq J(x_0) \geq \inf_{x \in D} J(x).$$

Then x_0 is the argument of a minimum of functional J over D. $\qquad\square$

In case $D = E$, the boundedness of any minimizing sequence is guaranteed when the sets J^α for each $\alpha \in \mathbb{R}$ are bounded. Such sets are bounded when the functional J is (weakly) coercive, i.e.,

$$\lim_{\|x\| \to +\infty} J(x) = +\infty.$$

In order to proceed further, we recall the differentiability notions required.

Definition 2.2   (*First Gâteaux variation*) Let $x_0 \in E$ be fixed. The functional $F : E \to \mathbb{R}$ has first Gâteaux variation at x_0 in direction $h \in E$ provided the following limit exists

$$\lim_{t \to 0} \frac{F(x_0 + th) - F(x_0)}{t} \tag{2.1}$$

that we denote by $\delta F(x_0, h)$.

Remark 2.2   Note that the limit in (2.1) is taken in $\mathbb{R}$ which means that the Gâteaux variation can be defined in case E is merely a linear space. Note also that the existence of the limit in the above definition is equivalent to the existence of the derivative at 0 of one variable real valued function $t \mapsto F(x_0 + th)$ with fixed $x_0, h \in E$.

Definition 2.3   (*Gâteaux derivative*) A functional $F : E \to \mathbb{R}$ is said to be Gâteaux differentiable at $x_0 \in E$ if there exists a bounded linear functional $F'(x_0) \in E^*$ such that for every $h \in E$

$$\lim_{t \to 0} \frac{F(x_0 + th) - F(x_0)}{t} = \langle F'(x_0), h \rangle. \tag{2.2}$$

The linear functional $F'(x_0)$ is then called the Gâteaux derivative of F at x_0.

Remark 2.3   Condition (2.2) in the definition of Gâteaux derivative is equivalent to saying that

$$F(x + th) - F(x) - t \langle F'(x_0), h \rangle = o(t),$$

where $\lim_{t \to 0} \frac{o(t)}{t} = 0$.

Remark 2.4   Note that the existence of the Gâteaux derivative does not imply neither continuity nor lower semicontinuity. Indeed, the function $F : \mathbb{R}^2 \to \mathbb{R}$ given by

$$F(x, y) = \begin{cases} 1, & x = y^2, y > 0 \\ 0, & \text{otherwise} \end{cases}$$

is Gâteaux differentiable at $(0, 0)$ while it is not continuous there.

Remark 2.5   The computation of the Gâteaux derivative is performed according to the steps:

(a) find the Gâteaux variation, i.e. calculate the limit in formula (2.2) for all points x in any direction h;

(b) prove that the *Gâteaux variation* is a bounded (and therefore continuous) linear mapping.

Now we proceed to some standard example about calculation of the Gâteaux derivative for Euler action functional. By 2^* we denote the Sobolev critical exponent for $N > 2$

$$2^* := \frac{2N}{N - 2}.$$

Note that $2^* > 2$.

Example 2.1 We put $E = H_0^1(\Omega)$ and fix $q \in (1, 2^*)$. Let $g : \mathbb{R} \to \mathbb{R}$ be a C^1 function such that there are constants $a, b > 0$

$$|g_u(u)| \leq a + b|u|^{q-1} \text{ for all } u \in \mathbb{R}. \tag{2.3}$$

Here g_u denotes the derivative of g. By a direct calculation we find numbers $a_1, b_1, c_1 > 0$ such that

$$|g(u)| \leq c_1 + b_1|u| + a_1|u|^q \text{ for all } u \in \mathbb{R}.$$

We consider the functional $F : E \to \mathbb{R}$ given by

$$F(x) = \int_\Omega g(u(x))\,dx + \frac{1}{2}\int_\Omega |\nabla u(x)|^2\,dx$$

which is well defined on E. We see that the second term is obviously C^1. We obtain that F has the Gâteaux derivative at every point u and in each direction h defined by the formula

$$\left\langle F'(u), h \right\rangle = \int_\Omega g_u(u(x))h(x)\,dx + \int_\Omega \nabla u(x)\,\nabla h(x)\,dx.$$

Indeed, the application of the Lagrange Mean Value Theorem says that

$$\left|\frac{g(z + \varepsilon w) - g(z)}{\varepsilon}\right| \leq \max_{c \in [z, w]} |g_u(c)|\,|w| \text{ for any } z, w \in \mathbb{R} \text{ and any } \varepsilon > 0.$$

By the above estimate together with assumption (2.3), we can apply *the Lebesgue Dominated Convergence Theorem* in order to differentiate under the integral sign. We easily check that

$$h \to \int_\Omega g_u(u(x))h(x)\,dx$$

defines a linear bounded functional on E. Thus we reach the conclusion that F is Gâteaux differentiable over E. In fact we can prove that F is continuously Gâteaux differentiable over E.

Necessary optimality condition reads now as follows:

Theorem 2.4 *Let $F : E \to \mathbb{R}$. If $x_0 \in E$ is a minimizer of F over E and if F is differentiable in the sense of Gâteaux at x_0, then for each $h \in E$ we have*

$$\left\langle F'(x_0), h \right\rangle = 0.$$

Now we are in a position to present the following version of the Weierstrass Theorem which is used to implement the so called direct method of the calculus of variations.

Theorem 2.5 (Direct method of the calculus of variations) *Let the functional $J :$ $E \to \mathbb{R}$ be Gâteaux differentiable, sequentially weakly lower semicontinuous and coercive. Then J has at least one minimizer x_0 over E,*

$$J(x_0) = \inf_{x \in E} J(x),$$

which is also a critical point, that is,

$$\langle J'(x_0), h \rangle = 0 \text{ for all } h \in E.$$

Proof Take such a number α that J^α is non-empty. Since the functional J is coercive it follows that J^α is bounded. We see at once that

$$\inf_{x \in E} J(x) = \inf_{x \in J^\alpha} J(x).$$

Using $D = J^\alpha$ we can apply Theorem 2.3 in order to find an element $x_0 \in J^\alpha$ such that $\inf_{x \in J^\alpha} J(x) = J(x_0)$. Hence x_0 is a minimizer of J over E. The application of Theorem 2.4 completes the proof. $\square$

We now proceed with some applications included into the examples that follow.

Example 2.2 We consider minimization of a classical action functional

$$J : H_0^1(\Omega) \to \mathbb{R}$$

defined by

$$J(u) = \frac{1}{2} \int_\Omega |\nabla u(x)|^2 \, dx + \frac{1}{\alpha} \int_\Omega |u(x)|^\alpha \, dx + \int_\Omega g(x) u(x) \, dx,$$

where $2 \le \alpha < 2^*$. Here $g \in L^2(\Omega)$. In case $N = 2$ we take any $\alpha > 1$.

We easily show that J is C^1 over $H_0^1(\Omega)$ and that

$$\left\langle F'(u), h \right\rangle = \int_\Omega \left(\nabla u(x) \nabla h(x) + |u(x)|^{\alpha-2} u(x) h(x) + g(x) h(x) \right) dx$$

for all $u, h \in H_0^1(\Omega)$. Let $N > 2$. When $\alpha = 2$, we see that the functional J is strictly convex. Moreover, for any $2 \le \alpha \le 2^*$ the functional J is coercive.

Example 2.3 Let $N > 2$. Let $g : \mathbb{R} \to \mathbb{R}$ be a continuously differentiable convex function with a derivative $g_u : \mathbb{R} \to \mathbb{R}$. We consider the functional $J : H_0^1(\Omega) \to \mathbb{R}$ given by

$$J(u) = \frac{1}{2} \int_\Omega |\nabla u(x)|^2 \, \mathrm{d}x + \int_\Omega g(u(x)) \, \mathrm{d}x.$$

Assume that there are constants $a, b > 0$ such that

$$|g_u(u)| \leq a + b \, |u|^{2^*-1} \text{ for all } u \in \mathbb{R}.$$

Then J is well defined, differentiable in the sense of Gâteaux, sequentially weakly lower semicontinuous and coercive. Since J is strictly convex, it has exactly one minimizer.

We can apply Theorem 2.5 in order to consider the solvability and the uniqueness of boundary value problems, We now follow with some standard example.

Example 2.4 We consider the following Dirichlet Problem (understood in a weak sense): find a function $u \in H_0^1(\Omega)$ such that the following equation is satisfied:

$$\begin{cases} -\Delta u(x) + g(x, u(x)) = 0, \\ \quad\quad u|_{\partial\Omega} = 0 \end{cases} \tag{2.4}$$

under the assumption that $N > 2$ and that
   $g : \Omega \times \mathbb{R} \to \mathbb{R}$ *is a Carathéodory function for which there is a function* $a \in L^2(\Omega)$ *and constant* $b > 0$ *and* $q \in (1, 2^*)$ *that*

$$|g(x, u)| \leq a + b \, |u|^{q-1} \text{ for all } u \in \mathbb{R} \text{ and a.e. } x \in \Omega. \tag{2.5}$$

We say that $u \in H_0^1(\Omega)$ is a weak solution to (2.4) provided that

$$\int_\Omega \nabla u(x) \nabla v(x) \, \mathrm{d}x + \int_\Omega g(x, u(x))v(x) \, \mathrm{d}x = 0 \text{ for all } v \in H_0^1(\Omega).$$

We define $G : \Omega \times \mathbb{R} \to \mathbb{R}$ by

$$G(x, u) = \int_0^u g(x, s) \, \mathrm{d}s \text{ for a.e. } x \in \Omega \text{ and all } u \in \mathbb{R}. \tag{2.6}$$

We can directly verify that G defined by (2.6) is a Carathéodory function and find that the relevant growth conditions (2.5) imposed on g lead to the conclusion that it is well defined over $H_0^1(\Omega)$. The corresponding Euler action functional $J : H_0^1(\Omega) \to \mathbb{R}$ (i.e. a functional whose critical points are weak solutions to (2.4)) is given by

$$J(u) = \frac{1}{2} \int_{\Omega} |\nabla u\,(x)|^2\,\mathrm{d}x + \int_{\Omega} G\,(x, u\,(x))\,\mathrm{d}x. \tag{2.7}$$

Under condition (2.5) the functional J given by (2.7) is well defined (i.e. finite for every $u \in H_0^1(\Omega)$), C^1 and sequentially weakly lower semicontinuous over $H_0^1(\Omega)$. For the sequential weak lower semicontinuity we use the properties of the norm which show that the term $u \mapsto \frac{1}{2} \int_{\Omega} |\nabla u\,(x)|^2\,\mathrm{d}x$ is sequentially weakly continuous (as convex and continuous) and also we utilize the compact embedding into $L^q\,(\Omega)$ for $q \in (1, 2^*)$ in order to investigate the second term. Indeed, given a weakly convergent sequence $\{u_n\}_{n \geq 1} \subset H_0^1(\Omega)$ with a weak limit u_0, we know that $\{u_n\}_{n \geq 1}$ converges strongly in $L^q\,(\Omega)$ to u_0. Then by the Lebesgue dominated convergence we see that

$$\lim_{n \to \infty} \int_{\Omega} G\,(x, u_n\,(x))\,\mathrm{d}x = \int_{\Omega} G\,(x, u_0\,(x))\,\mathrm{d}x.$$

In order to be able to apply Theorem 2.5 we need to assume a growth condition on G which leads to the coercivity of J, namely:

there exist functions $a \in L^{\infty}\,(\Omega)$, b, $c \in L^2\,(\Omega)$ and a number $\beta \in (0, 2)$ such that for a.e. $x \in \Omega$ and all $u \in \mathbb{R}$ it holds

$$G(x, u) \geq -a\,(x)\,|u|^{\beta} + b\,(x)\,u + c(x). \tag{2.8}$$

If we assume that conditions (2.5) and (2.8) are satisfied, then we show by a direct calculation that the functional J given by (2.7) is coercive. As a consequence from Theorem 2.5 we know that the Dirichlet Problem (2.4) has at least one weak solution $u_0 \in H_0^1(\Omega)$.

In case when G is convex we can alter condition (2.5) as follows:

$$|g\,(x, u)| \leq a + b\,|u|^{2^*-1} \textit{ for all } u \in \mathbb{R} \textit{ and a.e. } x \in \Omega. \tag{2.9}$$

This is due to the fact that upon convexity in order to prove that the functional J is sequentially weakly lower semicontinuous it suffices to demonstrate that it is well defined and continuous. Hence the compact embedding is not utilized. Moreover, we do not need condition (2.8) since in this case

$$G(x, u) \geq G\,(x, 0) + g\,(x, 0)\,u$$

for a.e. $x \in \Omega$ and all $u \in \mathbb{R}$. Concluding, in presence of convexity of G condition (2.9) suffices in order to get the existence and uniqueness of the Dirichlet Problem (2.4).

Various developments can be found in [10, 16, 22, 63, 74].

2.2 On the Ekeland Variational Principle

Another significant tool in critical point theory is the Ekeland Variational Principle. This principle serves as an alternative existence theorem for global minimizers of continuously differentiable functionals that are bounded from below, coercive, and satisfy the Palais-Smale condition. Specifically, it states that for any lower semicontinuous functional that is bounded from below and coercive, there exists a sequence that approximates the global minimum within any desired level of accuracy. The Palais-Smale condition, on the other hand, ensures that any sequence for which the functional is bounded and its derivative goes to zero. The Ekeland Variational Principle is widely used in various fields, including calculus of variations, optimization, and partial differential equations, due to its versatility and robustness.

Theorem 2.6 (Ekeland variational principle) *Assume $J : E \to \mathbb{R}$ is continuously Gâteaux differentiable and bounded from below. Then for any $\varepsilon > 0$ and $u \in E$ such that*

$$J(u) \leq \inf_{x \in E} J(x) + \varepsilon$$

there exists $v \in E$ such that:

(i) $J(v) \leq J(u)$,
(ii) $\|u - v\| \leq \sqrt{\varepsilon}$,
(iii) $\left\| J^{'}(v) \right\|_* \leq \sqrt{\varepsilon}$.

For the proof we refer to [21].

We introduce the following notion regarding a sequence of almost critical points:

Definition 2.4 (*Palais-Smale condition*) Assume that $J : E \to \mathbb{R}$ is differentiable in the sense of Gâteaux. We say that J satisfies the Palais-Smale condition, (PS)-condition for short, if for every sequence $\{x_n\}_{n \geq 1} \subset E$ such that

(i) the sequence $\{J(x_n)\}_{n \geq 1}$ is bounded,
(ii) $J'(x_n) \to 0$ in E^*

has a norm convergent subsequence.

Remark 2.6 Any functional on a finite dimensional normed space which is coercive and differentiable in the sense of Gâteaux, satisfies the (PS)-condition.

With the aid of the Ekeland Variational Principle and the (PS)-condition we can derive a sort of a counterpart of Theorem 2.5 when the assumption of the sequential weak lower semicontinuity is dropped. The assumption that $J : E \to \mathbb{R}$ is continuously differentiable is also described as J being a C^1 functional.

Theorem 2.7 (Existence for a C^1 functional) *Assume that $J : E \to \mathbb{R}$ is continuously differentiable, bounded from below and satisfies the (PS)-condition. Then there exists $x_0 \in E$ such that*

$$J(x_0) = \inf_{x \in E} J(x).$$

Related results can be found in [21, 23, 39, 44, 57, 65, 71].

2.3 Monotone Methods

The method of monotone operators is central in our reasoning. For the understanding of what follows, it is necessary to become acquainted with several related notions, tools, and techniques that we just introduced and commented on. Some proofs are included that will further influence the reasoning connected to competing operators. Additional material is available in [7, 27, 30, 44].

Definition 2.5 (*Various types of monotonicity*) The operator $A : E \to E^*$ is called:

(i) monotone, if for all $u, v \in E$ it holds

$$\langle A(u) - A(v), u - v \rangle \geq 0;$$

(ii) d-monotone, if for some increasing function $\rho : [0, +\infty) \to \mathbb{R}$, it holds for all $u, v \in E$,

$$\langle A(u) - A(v), u - v \rangle \geq (\rho(\|u\|) - \rho(\|v\|))(\|u\| - \|v\|);$$

(iii) radially continuous, if for all $u, v \in E$ the function

$$s \to \langle A(u + sv), v \rangle$$

is continuous on $\mathbb{R}$;

(iv) hemicontinuous, if for all $u, v, h \in E$ the function

$$s \to \langle A(u + sv), h \rangle$$

is continuous on $\mathbb{R}$;

(v) demicontinuous if $u_n \to u_0$ in E implies $A(u_n) \rightharpoonup A(u_0)$ in E^*;

(vi) strongly continuous if $u_n \rightharpoonup u_0$ in E implies $A(u_n) \to A(u_0)$ in E^*;

(vii) weakly continuous if $u_n \rightharpoonup u_0$ in E implies $A(u_n) \rightharpoonup A(u_0)$ in E^*;

(viii) weakly coercive when

$$\lim_{\|u\| \to +\infty} \|A(u)\|_* = +\infty;$$

(ix) coercive if

$$\lim_{\|u\| \to +\infty} \frac{\langle A(u), u \rangle}{\|u\|} = +\infty.$$

Here an operator sends points from E into functionals from E^*. Continuity of the operator A is understood as usual, i.e., for all sequences $u_n \to u_0$ in E, we have $A(u_n) \to A(u_0)$ in E^*. In case of problems considered in Sobolev spaces, the strong continuity can be checked with the help of a compact embedding and the Krasnosel'skii Theorem. A strongly continuous operator is compact (i.e., it is continuous and sends bounded sets into relatively compact ones). Note that if there exists a function

$$\gamma : [0, +\infty) \to \mathbb{R}, \text{ with } \lim_{x \to +\infty} \gamma(x) = +\infty$$

such that
$$\langle A(u), u \rangle \geq \gamma(\|u\|) \|u\| \text{ for all } u \in E,$$

then A is coercive. We proceed now to some example of a very general monotone mapping which concerns several applications mentioned further on. Let $p \geq 2$ and let Ω be an open subset of $\mathbb{R}^N$. We assume that

Aφ $\varphi : \Omega \times \mathbb{R}_+ \to \mathbb{R}$ is a Carathéodory function for which there is a constant $M > 0$ such that

$$|\varphi(x, u)| \leq M \text{ for a.e. } x \in \Omega \text{ and all } u \in \mathbb{R}_+.$$

Under some additional growth assumption on function φ we will consider the monotonicity of the operator

$$A : L^p(\Omega) \to L^{p'}(\Omega),$$

given by

$$\langle A(u), v \rangle = \int_{\Omega} \varphi\left(x, |u(x)|^{p-1}\right) |u(x)|^{p-2} u(x) v(x) \, dx \qquad (2.10)$$

or else a.e. on Ω

$$A(u(\cdot)) = \varphi\left(\cdot, |u(\cdot)|^{p-1}\right) |u(\cdot)|^{p-2} u(\cdot).$$

We verify by a direct calculation that A is well defined and that it is continuous which fact follows by the application of the Krasnosel'skii Theorem.

Theorem 2.8 (Conditions for monotonicity) *Assume that condition Aφ is satisfied. The following assertions hold.*

    (i) *If*

$$\varphi(x, u) u - \varphi(x, v) v \geq 0$$

for all $u \geq v \geq 0$ and a.e. $x \in \Omega$, then A is monotone.

(ii) *If there exists a constant $\gamma > 0$ such that*

$$\varphi(x, u)\, u - \varphi(x, v)\, v \geq \gamma\, (u - v)$$

for all $u \geq v \geq 0$ and a.e. $x \in \Omega$, then A is d-monotone with respect to the function

$$\rho(x) = \gamma x^{p-1}.$$

(iii) *If for some $m > 0$*

$$\varphi(x, u) \geq m$$

for all $u \in \mathbb{R}_+$ and a.e. $x \in \Omega$, then A is coercive.

(iv) *Let $p = 2$ and let there exist a constant $\gamma > 0$ such that*

$$\varphi(x, u)\, u - \varphi(x, v)\, v \geq \gamma\, (u - v) \tag{2.11}$$

for all $u \geq v \geq 0$ and a.e. $x \in \Omega$. Then A is strongly monotone.

Proof We provide the sketch of the proof. For more details we suggest to consult [31]. Firstly we prove (i). Observe that for $u, v \in L^p(\Omega)$

$$\langle A(u) - A(v), u - v \rangle$$

$$= \int_\Omega \varphi\left(x, |u|^{p-1}\right) |u|^{p-2} u\, (u - v)\, \mathrm{d}x - \int_\Omega \varphi\left(x, |v|^{p-1}\right) |v|^{p-2} v\, (u - v)\, \mathrm{d}x$$

$$\geq \int_\Omega \left(\varphi\left(x, |u|^{p-1}\right) |u|^{p-1} - \varphi\left(x, |v|^{p-1}\right) |v|^{p-1}\right) (|u| - |v|)\, \mathrm{d}x \geq 0.$$

(ii). From (i) and by the Hölder Inequality we see for $u, v \in L^p(\Omega)$

$$\langle A(u) - A(v), u - v \rangle$$

$$\geq \int_\Omega \left(\varphi\left(x, |u|^{p-1}\right) |u|^{p-1} - \varphi\left(x, |v|^{p-1}\right) |v|^{p-1}\right) (|u| - |v|)\, \mathrm{d}x$$

$$\geq \gamma \int_\Omega |u|^p\, \mathrm{d}x + \gamma \int_\Omega |v|^p\, \mathrm{d}x - \gamma \int_\Omega \left(|u|^{p-1}|v| - |v|^{p-1}|u|\right) \mathrm{d}x$$

$$\geq \gamma\, (\|u\|_{L^p} - \|v\|_{L^p}) \left(\|u\|_{L^p}^{p-1} - \|v\|_{L^p}^{p-1}\right).$$

(iii). Note that for any $u \in L^p(\Omega)$ we have

$$\langle A(u), u \rangle = \int_\Omega \varphi\left(x, |u|^{p-1}\right) |u|^p\, \mathrm{d}x \geq m\, \|u\|_{L^p}^p.$$

This estimate implies the coercivity.

(iv). Since (2.11) holds we can define the function φ_0 by

$$\varphi_0 \left(u \left(\cdot \right) \right) = \varphi \left(\cdot, \left| u \left(\cdot \right) \right| \right) - \gamma \text{ for } u \in L^2 \left(\Omega \right).$$

Denote by A_0 the operator defined by (2.10) with φ_0 instead of φ. Then $\varphi = \varphi_0 + \gamma$. We also put $\varphi_1 \left(u \left(\cdot \right) \right) = \gamma$ for $u \in L^2 \left(\Omega \right)$ and observe that operator

$$A_1 : L^2 \left(\Omega \right) \to L^2 \left(\Omega \right)$$

determined by φ_1 according to formula (2.10) with φ_1 instead of φ is strongly monotone. By part (i) it follows that operator A_0 is monotone. A sum of a strongly monotone and a monotone operator, that is operator A, is necessarily strongly monotone. $\square$

By the Banach-Steinhaus Theorem we reach the following boundedness property which will be essential in getting that the Galerkin type approximations are bounded as we will see in the proof of Minty-Browder Theorem:

Proposition 2.1 *Let $A : E \to E^*$ be a monotone operator. Then A is locally bounded, which means that for each fixed $u \in E$ there are constants $\varepsilon > 0$ and $M > 0$ such that it holds $\| A \left(v \right) \|_* \leq M$ for all v satisfying $\| u - v \| \leq \varepsilon$.*

Remark 2.7 Note that if $A : E \to E^*$ is demicontinuous, then it is hemicontinuous and locally bounded. (Observe that monotonicity is not assumed).

Definition 2.6 (*Bounded operator*) An operator $A : E \to E^*$ is called bounded when for any bounded set $B \subset E$, the set $A \left(B \right) \subset E^*$ is bounded.

Recall that a linear operator is bounded if and only it is continuous. A (non-linear) continuous operator need not be bounded as seen from the following example:

Example 2.5 Let us consider the continuous operator $A : l^2 \to l^2$ given by

$$A \left(u \right) = \left(u_1, u_2^2, u_3^4, \dots \right).$$

Set

$$e_n = \{ \delta_{in} \}_{i \geq 1}, \text{ for } n \in \mathbb{N},$$

where

$$\delta_{in} = \begin{cases} 1, & i = n, \\ 0, & i \neq n \end{cases}$$

for $i, n \in \mathbb{N}$. For the sequence $\{ 2e_n \}_{n \geq 1} \subset l^2$ we have $\| 2e_n \| = 2$ and $\| A \left(2e_n \right) \| = 2^n$. Thus A is not bounded.

The Krasnosel'skii Theorem, see Theorem 1.1 about the continuity of the composition operators helps us in proving that a given nonlinear operator is bounded. Note that a monotone operator need not be bounded.

Example 2.6 The operator $A : l^2 \to l^2$ given by

$$A(x) = (f_1(x_1), f_2(x_2), ...) + (x_1, x_2, ...),$$

where

$$f_n(t) = \begin{cases} 0, & \text{for } t \le \frac{1}{2}, \\ nt - \frac{n}{2}, & \text{for } t > \frac{1}{2}, \end{cases}$$

fulfills that the image $A(B)$ of the unit ball B is unbounded.

We complete this short introduction on monotonicity with some compactness conditions to be used further on. Recall that in a uniformly convex space relations $u_n \rightharpoonup u_0$ and $\|u_n\| \to \|u_0\|$ imply that $u_n \to u_0$. This suggests the following notion:

Definition 2.7 (*Condition (S)*) The operator $A : E \to E^*$ satisfies condition (S) if the relations

$$u_n \rightharpoonup u_0 \text{ in } E$$

and

$$\langle A(u_n) - A(u_0), u_n - u_0 \rangle \to 0$$

imply that

$$u_n \to u_0 \text{ in } E.$$

Lemma 2.2 *Assume that E is a uniformly convex space. Let $A : E \to E^*$ be $d-$ monotone. Then A satisfies condition (S).*

Proof Assume that $u_n \rightharpoonup u_0$ in E and that $\langle A(u_n) - A(u_0), u_n - u_0 \rangle \to 0$. Then the property of A to be $d-$ monotone, see Definition 2.5, entails that $\|u_n\| \to \|u_0\|$. Since the space E is uniformly convex, we conclude that $u_n \to u_0$ in E, which completes the proof. $\qquad\qquad\square$

Lemma 2.3 *Assume that the operator $A : E \to E^*$ fulfills property (S) and that $T : E \to E^*$ is strongly continuous. Then $A + T$ also has property (S).*

Another applicable condition called $(S)_2$, which is weaker than condition (S), is stated below.

Definition 2.8 (*Condition $(S)_2$*) The operator $A : E \to E^*$ satisfies condition $(S)_2$ if $u_n \rightharpoonup u_0$ in E and $A(u_n) \to A(u_0)$ in E^* imply that $u_n \to u_0$ in E.

Other conditions making weakly convergent sequences strongly convergent, via a suitable action of operator, are as follows:

Condition $(S)_+$:

$$u_n \rightharpoonup u_0 \quad in \quad E \quad and \quad \limsup_{n \to +\infty} \langle A(u_n) - A(u_0), u_n - u_0 \rangle \le 0$$

imply that $u_n \to u_0$ in E;

Condition $(S)_0$:

$$u_n \rightharpoonup u_0, \ A(u_n) \rightharpoonup b, \ \langle A(u_n), u_n \rangle \to \langle b, u_0 \rangle$$

imply that $u_n \to u_0$ in E.

Remark 2.8 In the definitions of conditions (S) and $(S)_+$ one can replace

$$\langle A(u_n) - A(u_0), u_n - u_0 \rangle \to 0$$

with

$$\langle A(u_n), u_n - u_0 \rangle \to 0$$

since when $u_n \rightharpoonup u_0$ in E we have

$$\langle A(u_0), u_n - u_0 \rangle \to 0.$$

If the condition $(S)_+$ holds, condition (S) is satisfied and this in turn implies the condition $(S)_0$. We mostly use the condition $(S)_+$ in what follows. Since most operators satisfying $(S)_+$ also fulfill the condition S, sometimes this assumption will be imposed.

The concepts that we introduce next allow for connecting monotonicity methods with variational ones.

Definition 2.9 (*Potential operator*) The operator $A : E \to E^*$ is potential, if there exists a functional $F : E \to \mathbb{R}$ which is differentiable in the sense of Gâteaux on E and is such that $F' = A$. The functional F is called the potential of A.

Example 2.7 Given $p \in (1, \infty)$, the negative p-Laplacian

$$-\Delta_p : W_0^{1,p}(\Omega) \to W^{-1,p'}(\Omega),$$

that is,

$$\langle -\Delta_p u, v \rangle := \int_\Omega |\nabla u(x)|^{p-2} \nabla u(x) \nabla v(x) \, dx \ \text{for} \ u, v \in W_0^{1,p}(\Omega)$$

is a potential operator whose potential $F : W_0^{1,p}(\Omega) \to \mathbb{R}$ is

$$F(u) = \frac{1}{p} \int_\Omega |\nabla u(x)|^p \, dx = \frac{1}{p} \|u\|_{W_0^{1,p}(\Omega)}^p, \quad \forall u \in W_0^{1,p}(\Omega).$$

In case $p = 2$ we recover the negative Laplacian. From Theorem 2.8 we have that the negative $p-$ Laplacian is d-monotone with respect to the function $\rho(x) = x^{p-1}$

and the negative Laplacian is strongly monotone (in the case when $p = 2$). These operators are bounded and continuous due to the remarks preceding Theorem 2.8.

Theorem 2.9 *Assume that condition $A\varphi$ is satisfied. The operator A given by (2.10) is potential with the potential $F : L^p(\Omega) \to \mathbb{R}$ defined by*

$$F(u) = \int_\Omega \int_0^{|u(x)|} \varphi\left(x, s^{p-1}\right) s^{p-1} \mathrm{d}s \mathrm{d}x \ \textit{for } u \in L^p(\Omega).$$

Proof For any fixed $u, v \in L^p(\Omega)$ we have by a direct differentiation

$$\lim_{\lambda \to 0} \frac{F(u+\lambda v) - F(u)}{\lambda} = \frac{\mathrm{d}}{\mathrm{d}\lambda} \int_\Omega \int_0^{|u(x)+\lambda v(x)|} \varphi\left(x, s^{p-1}\right) s^{p-1} \mathrm{d}s \mathrm{d}x \bigg|_{\lambda=0}$$

$$= \int_\Omega \varphi\left(x, |u(x)|^{p-1}\right) |u(x)|^{p-2} u(x) v(x) \mathrm{d}x = \langle A(u), v \rangle.$$

$\square$

Remark 2.9 If the operator $A : E \to E^*$ is potential and demicontinuous, then for any $v \in E$ we can find the formula for its potential:

$$F(v) = F(0) + \int_0^1 \langle A(sv), v \rangle \mathrm{d}s.$$

Remark 2.10 Observe that if the operator A is potential, then the Gâteaux differentiability of the potential F is guaranteed. If A is continuous, then F is a C^1 functional. On the other hand, the Gâteaux differentiability of a functional does not imply its continuity. The next result provides a sort of converse of the monotonicity of the derivative of a convex functional. If $A : E \to E^*$ is potential and monotone, then its potential $F : E \to \mathbb{R}$ is convex and sequentially weakly lower semicontinuous. The monotone operators may lack any suitable type of continuity, even in finite-dimensional spaces. This cannot happen for monotone and potential operators since a potential and monotone operator is demicontinuous. Moreover, if $A : E \to E^*$ is potential, demicontinuous, bounded and coercive, then its potential $F : E \to \mathbb{R}$ is coercive.

2.4   On the (Degenerate) p-Laplacian and (p, q)-Laplacian

Now we outline some basic facts about the p-Laplacian operator with $p \in (1, +\infty)$ and its weighted counterpart. The (negative) degenerate p-Laplacian with the nonnegative weight $a \in L^1(\Omega)$ is the map

$$- \Delta_p^a : W_0^{1,p}(a, \Omega) \to W_0^{1,p}(a, \Omega)^*$$

defined by

$$\left\langle -\Delta_p^a(u), v \right\rangle := \int_\Omega a(x)|\nabla u|^{p-2}\nabla u \nabla v\, dx, \quad \forall u, v \in W_0^{1,p}(a, \Omega).$$

It is readily seen that the definition of $-\Delta_p^a$ makes sense since by Höder's inequality it holds

$$\left| \int_\Omega a(x)|\nabla u(x)|^{p-2}\nabla u(x)\nabla v(x) dx \right|$$

$$\leq \left(\int_\Omega a(x)|\nabla u(x)|^p dx \right)^{\frac{p-1}{p}} \left(\int_\Omega a(x)|\nabla v(x)|^p dx \right)^{\frac{1}{p}} < +\infty$$

for all $u, v \in W_0^{1,p}(a, \Omega)$. The operator $-\Delta_p^a : W_0^{1,p}(a, \Omega) \to W_0^{1,p}(a, \Omega)^*$ is continuous and bounded (in the sense that it maps bounded sets into bounded sets).

When the nonnegative weight a is nontrivial, it possesses a first eigenvalue λ_1^a, that is the least one $\lambda > 0$ for which the problem

$$\begin{cases} -\Delta_p^a u = \lambda|u|^{p-2}u & \text{in } \Omega \\ u = 0 & \text{on } \partial\Omega, \end{cases} \tag{2.12}$$

possesses a nontrivial solution. From (2.12), the first eigenvalue λ_1^a is given by

$$\lambda_1^a := \inf_{u \in W_0^{1,p}(a,\Omega), u \neq 0} \frac{\int_\Omega a(x)|\nabla u|^p dx}{\int_\Omega |u|^p dx}.$$

In particular, the (negative) p-Laplacian $-\Delta_p : W_0^{1,p}(\Omega) \to W^{-1,p'}(\Omega)$ is obtained for $a(x) \equiv 1$, thus

$$\left\langle -\Delta_p(u), v \right\rangle = \int_\Omega |\nabla u|^{p-2}\nabla u \nabla v\, dx, \quad \forall u, v \in W_0^{1,p}(\Omega).$$

The first eigenvalue of $-\Delta_p$ is denoted λ_1, so

$$\lambda_1 := \inf_{u \in W_0^{1,p}(\Omega), u \neq 0} \frac{\int_\Omega |\nabla u|^p dx}{\int_\Omega |u|^p dx}.$$

The ordinary (negative) Laplacian $-\Delta$ is obtained when $p = 2$. We refer to [50] for the study of the p-Laplacian.

We pass to the (degenerate) (p, q)-Laplacian. Let $1 < q < p < +\infty$ and the nonnegative weights $a, b \in L^1(\Omega)$ satisfying condition (1.5). By Proposition 1.1 the operator

$$- \Delta_{p,q}^{a,b} : W_0^{1,p}(a, \Omega) \to W_0^{1,p}(a, \Omega)^*$$

given by

$$\langle -\Delta_{p,q}^{a,b} u, \, v \rangle_{W_0^{1,p}(a,\Omega)} \tag{2.13}$$

$$:= \int_\Omega (a(x)|\nabla u|^{p-2}\nabla u + b(x)|\nabla u|^{q-2}\nabla u)\nabla v(x)\mathrm{d}x$$

for all $u, v \in W_0^{1,p}(a, \Omega)$ is well-defined. The operator $-\Delta_{p,q}^{a,b}$ is called the (negative) degenerate (p, q)-Laplacian with weights $a, b \in L^1(\Omega)$. A direct verification shows that:

Lemma 2.4 (Monotonicity properties of the degenerate *(p, q)*-Laplacian) *The operator $-\Delta_{p,q}^{a,b}$ is continuous, strictly monotone and bounded.*

When no weights appear, we recover the usual (p, q)-Laplacian. The (negative) degenerate p-Laplacian

$$- \Delta_p^a : W_0^{1,p}(a, \Omega) \to W_0^{1,p}(a, \Omega)^*$$

with weight $a \in L^1(\Omega)$ is recovered when $b = 0$. The (negative) (p, q)-Laplacian, i.e., $-\Delta_p - \Delta_q$, is retrieved for $a = b \equiv 1$. If $a \equiv 1$ and $b \in L^\infty(\Omega)$, we get the driving operator in [42, 60]. When $b \equiv 1$, the degenerate (p, q)-Laplacian $-\Delta_{p,q}^{a,b} : W_0^{1,p}(a, \Omega) \to W^{-1,p'}(\Omega)$ becomes the double phase operator (see [37]). We close this section with an important property of this class of operators. We will return to related properties later on, when we will discuss background on monotonicity methods.

Proposition 2.2 *Assume that hypothesis (H) and condition (1.5) hold. Then the $(S)_+$ property is fulfilled for the operator $-\Delta_{p,q}^{a,b} : W_0^{1,p}(a, \Omega) \to W_0^{1,p}(a, \Omega)^*$ defined in (2.13), which means that every sequence $\{u_n\}_{n\geq 1} \subset W_0^{1,p}(a, \Omega)$ satisfying $u_n \rightharpoonup u$ in $W_0^{1,p}(a, \Omega)$ and*

$$\limsup_{n\to\infty}\langle -\Delta_{p,q}^{a,b}(u_n), u_n - u \rangle_{W_0^{1,p}(a,\Omega)} \leq 0 \tag{2.14}$$

verifies $u_n \to u$ in $W_0^{1,p}(a, \Omega)$.

Proof Let a sequence $\{u_n\}_{n\geq 1} \subset W_0^{1,p}(a, \Omega)$ be as in the statement above. By the monotonicity of $-\Delta_q^b$ and Hölder's inequality, we get

$$
\begin{aligned}
&\langle -\Delta_{p,q}^{a,b}(u_n) + \Delta_{p,q}^{a,b}(u), u_n - u \rangle_{W_0^{1,p}(a,\Omega)} \\
&\geq \langle -\Delta_p^a(u_n) + \Delta_p^a(u), u_n - u \rangle_{W_0^{1,p}(a,\Omega)} \\
&\geq (\|u_n\|_{W_0^{1,p}(a,\Omega)} - \|u\|_{W_0^{1,p}(a,\Omega)})(\|u_n\|_{W_0^{1,p}(a,\Omega)}^{p-1} - \|u\|_{W_0^{1,p}(a,\Omega)}^{p-1}) \geq 0.
\end{aligned}
$$

Due to (2.14), the preceding estimate results in

$$
\lim_{n \to +\infty} \|u_n\|_{W_0^{1,p}(a,\Omega)} = \|u\|_{W_0^{1,p}(a,\Omega)}.
$$

It is known from Proposition 1.1 that the space $W_0^{1,p}(a, \Omega)$ is uniformly convex. This implies that $u_n \to u$ in $W_0^{1,p}(a, \Omega)$. The proof is complete. $\qquad\square$

Corollary 2.1 *Assume that hypothesis (H) holds. Then the $(S)_+$ property is fulfilled for the operator $-\Delta_p^a : W_0^{1,p}(a, \Omega) \to W_0^{1,p}(a, \Omega)^*$ meaning that every sequence $\{u_n\} \subset W_0^{1,p}(a, \Omega)$ that satisfies $u_n \rightharpoonup u$ in $W_0^{1,p}(a, \Omega)$ and*

$$
\limsup_{n\to\infty} \langle -\Delta_p^a(u_n), u_n - u \rangle_{W_0^{1,p}(a,\Omega)} \leq 0
$$

verifies $u_n \to u$ in $W_0^{1,p}(a, \Omega)$. In particular, the (negative) p-Laplacian $-\Delta_p : W_0^{1,p}(\Omega) \to W_0^{1,p}(\Omega)^*$ satisfies the $(S)_+$ property. Moreover, for $1 < q < p < +\infty$, the (negative) (p, q)-Laplacian $-\Delta_p - \Delta_q : W_0^{1,p}(\Omega) \to W_0^{1,p}(\Omega)^*$ satisfies the $(S)_+$ property meaning that every sequence $\{u_n\} \subset W_0^{1,p}(\Omega)$ that satisfies $u_n \rightharpoonup u$ in $W_0^{1,p}(\Omega)$ and

$$
\limsup_{n\to\infty} \left[\langle -\Delta_p(u_n), u_n - u \rangle_{W_0^{1,p}(\Omega)} + \langle -\Delta_q(u_n), u_n - u \rangle_{W_0^{1,q}(\Omega)} \right] \leq 0
$$

fulfils $u_n \to u$ in $W_0^{1,p}(\Omega)$.

Proof For the first part, it suffices to take $b(x) \equiv 0$ in Proposition 2.2, while for the last part, $a(x) = b(x) \equiv 1$. $\qquad\square$

2.5 On the (Negative) Variable $p(x)$-Laplacian

Let $p \in C(\bar{\Omega})$ such that $p(x) > 1$ for all $x \in \bar{\Omega}$. As in the case of the (negative) p-Laplacian, we introduce the operator

$$
-\Delta_{p(x)} : W_0^{1,p(x)}(\Omega) \to \left(W_0^{1,p(x)}(\Omega) \right)^*
$$

defined by

$$\langle -\Delta_{p(x)}u, h\rangle := \int_\Omega |\nabla u(x)|^{p(x)-2}\nabla u(x) \cdot \nabla h(x)\,dx \quad \text{for all } u, h \in W_0^{1,p(x)}(\Omega),$$

and, for $q \in C(\bar\Omega)$ such that $q(x) > 1$ for all $x \in \bar\Omega$, the operator

$$\Delta_{q(x)} : W_0^{1,q(x)}(\Omega) \to \left(W_0^{1,q(x)}(\Omega)\right)^*$$

defined by

$$\langle \Delta_{q(x)}u, h\rangle := -\int_\Omega |\nabla u(x)|^{q(x)-2}\nabla u(x) \cdot \nabla h(x)\,dx \quad \text{for all } u, h \in W_0^{1,q(x)}(\Omega).$$

Both operators $-\Delta_{p(x)}$ and $\Delta_{q(x)}$ represent, respectively, the (negative) variable Laplacian differential operator and the positive $q(x)$-Laplace differential operator. In the following lemma, we summarize the properties of the $-\Delta_{p(x)}$-Laplacian. We emphasize this approach, as it appears to be novel in the literature. To facilitate the proof, we include some abstract background directly, following [31].

Assume that E is a real, separable, and reflexive Banach space. We say that the operator $A : E \to E^*$ is *uniformly monotone* if there exists an increasing function

$$r : [0, +\infty) \to [0, +\infty) \text{ with } r(0) = 0$$

and such that for all $u, v \in E$,

$$\langle A(u) - A(v), u - v\rangle \geq r\left(\|u - v\|\right)\|u - v\|.$$

Lemma 2.5 *Assume that the operator $A : E \to E^*$ is uniformly monotone. Then it is strictly monotone, coercive, and of type $(S)_+$.*

In view of the above lemma, it suffices to prove that the $-\Delta_{p(x)}$-Laplacian is uniformly monotone in order to establish its other important properties. Continuity of this operator is obvious.

Lemma 2.6 *The $-\Delta_{p(x)}$-Laplacian operator is continuous and uniformly monotone.*

Proof For any $u, v \in W_0^{1,p(x)}(\Omega)$, following Remark 1.3, we observe that

$$\langle -\Delta_{p(x)}u - (-\Delta_{p(x)}v), u - v\rangle = \int_\Omega |\nabla u(x) - \nabla v(x)|^{p(x)}\,dx$$

$$\geq r\left(\|\nabla u - \nabla v\|_{p(x)}\right)\|\nabla u - \nabla v\|_{p(x)},$$

where the function r is defined by Eq. (1.6). This proves the assertion. $\qquad\square$

From the above lemma and known results in [64], we conclude:

Lemma 2.7 *The $-\Delta_{p(x)}$-Laplacian operator is linear bounded, continuous, coercive, and uniformly monotone, and hence strictly monotone and of type $(S)_+$.*

The $\Delta_{q(x)}$-Laplacian is continuous and bounded.

2.6 On the (Negative) s-Fractional p-Laplacian

Suppose that Ω is a bounded open subset of $\mathbb{R}^N$ with Lipschitz boundary and we let $p \in (1, \infty)$. In order to obtain a fractional version of the (negative) p-Laplacian, we suppose that $s \in (0, 1)$ and let

$$X(\Omega) := \left\{ u \in W_p^s\left(\mathbb{R}^N\right) : u = 0 \text{ a.e. in } \mathbb{R}^N \backslash \Omega \right\}.$$

The space $X(\Omega)$ is endowed with the norm $[\cdot]_{s,p,\mathbb{R}^N}$ and then becomes a uniformly convex Banach space. We will keep standard notation $W_0^{s,p}(\Omega)$ for this space. The nonlinear map $A : W_0^{s,p}(\Omega) \to W^{-s,p'}(\Omega)$ defined by

$$\langle A(u), v \rangle = \int\limits_{\mathbb{R}^N} \int\limits_{\mathbb{R}^N} \frac{|u(x) - u(y)|^{p-2}(u(x) - u(y))(v(x) - v(y))}{|x - y|^{N+sp}} dx dy$$

for all $u, v \in W_0^{s,p}(\Omega)$ is a duality map and it is called the s-fractional p-Laplacian and denoted by $(-\Delta)_p^s$; it is the gradient of the functional $J : W_0^{s,p}(\Omega) \to \mathbb{R}$ defined by

$$J(u) = \frac{1}{p} \int\limits_{\mathbb{R}^N} \int\limits_{\mathbb{R}^N} \frac{|u(x) - u(y)|^p}{|x - y|^{N+sp}} dx dy = \frac{1}{p}[u]_{s,p,\mathbb{R}^N}^p$$

and therefore the fractional p-Laplacian is potential. In order to investigate further properties we recall the following standard inequality. For all $a, b \in \mathbb{R}$:

$$\left(|b|^{p-2}b - |a|^{p-2}a\right)(b - a) \geq \begin{cases} (p-1)|b - a|^2 \left(|a|^2 + |b|^2\right)^{-(2-p)/2} & \text{if } p \in (1, 2], \\ 2^{2-p}|b - a|^p & \text{if } p \in (2, \infty). \end{cases}$$

Utilizing the above inequality we prove, as in the classical case, that the operator $(-\Delta)_p^s$ is d-monotone and since the space $W_0^{s,p}(\Omega)$ is uniformly convex, we see that $(-\Delta)_p^s$ is of type $(S)_+$. Summarizing, we have

Lemma 2.8 *The map $(-\Delta)_p^s : W_0^{s,p}(\Omega) \to W^{-s,p'}(\Omega)$ is continuous, d-monotone (and thus a homeomorphism) and of type $(S)_+$.*

The first eigenvalue of the s-fractional p-Laplacian is expressed as

$$\lambda_{1,p,s} := \inf_{\substack{u \in W_0^{s,p}(\Omega) \\ u \neq 0}} \frac{\|u\|_{s,p}^p}{\|u\|_p^p}.$$

For more involved information about the s-fractional p-Laplacian and associated boundary-value problems we refer to [4, 5, 40, 41].

2.7   The Browder-Minty Theorem

In this section we will set forth the main existence tool proven via finite dimensional approximations. The proof of the Browder-Minty Theorem relies on the analysis of the convergence of the so called Galerkin type approximations. Some technical lemmas are given in order to provide suitable background for the proof of the main existence result.

Lemma 2.9 *Assume that $A : E \to E^*$ is coercive. Then for any $f \in E^*$ the set K of solutions to the equation $A(u) = f$ is bounded.*

Lemma 2.10 *Assume that $A : E \to E^*$ is radially continuous and monotone. Then for any $f \in E^*$ the set K of solutions to the equation $A(u) = f$ is sequentially weakly closed and convex.*

Lemma 2.11 *Assume that the operator $A : E \to E^*$ is monotone and $K \subset E$ is such a set that*

$$\|u\| \leq M_1 \text{ and } \langle A(u), u \rangle \leq M_2 \text{ for each } u \in K,$$

where M_1, M_2 are constants. Then there exists a constant M such that

$$\|A(u)\|_* \leq M \text{ for all } u \in K.$$

Note that the above lemma follows without monotonicity if we assume the operator A to be bounded.

Remark 2.11 Since E is separable it contains a dense and countable set $\{h_1, ..., h_n, ...\}$. Define E_n for $n \in \mathbb{N}$ as the linear hull of $\{h_1, ..., h_n\}$. The sequence of subspaces E_n has the approximation property: for each $u \in E$ there is a sequence $\{u_n\}_{n \geq 1}$ such that $u_n \in E_n$ for $n \in \mathbb{N}$ and $u_n \to u$. Such a sequence of subspaces is referred to as a **Galerkin base**.

For the proof of the main existence result related to equations involving monotone mappings we need the Brouwer Fixed Point Theorem and one of its corollaries. As it was with the case of the Weierstrass-Tonelli Theorem, when proving the Browder-Minty Theorem, we approximate a solution with a suitably chosen sequence. Its boundedness is reached via the coercivity and convergence via monotonicity and appropriate continuity.

Theorem 2.10 (Brouwer fixed point theorem) *Let $C \subset \mathbb{R}^N$ be a closed, bounded and convex set. Suppose that $f : \mathbb{R}^N \to \mathbb{R}^N$ is continuous and that $f(C) \subseteq C$. Then f has a fixed point in C, i.e., there is $u \in C$ such that*

$$f(u) = u.$$

The following consequence of Brouwer Fixed Point Theorem will be a useful tool in the sequel. Its proof is usually done on $\mathbb{R}^N$ endowed with the standard Euclidean norm (see, e.g., [73]). Here, on the pattern of [46], we prove it on an arbitrary finite dimensional space V without claiming that it directly follows from the identification of V with $\mathbb{R}^N$.

Lemma 2.12 *Let V be a finite dimensional real Banach space endowed with the norm $\|\cdot\|_V$ and let $F : V \to V^*$ be a continuous map. Assume that there is a constant $R > 0$ such that*

$$\langle F(v), v \rangle \geq 0 \text{ for all } v \in V \text{ with } \|v\|_V = R. \tag{2.15}$$

Then there exists $u \in V$ with $\|u\|_V \leq R$ satisfying

$$F(u) = 0.$$

Proof Let N be the dimension of V and fix a basis $\{v_1, ..., v_N\}$ of V. Suppose by contradiction that $F(v) \neq 0$ for all $\|v\|_V \leq R$. We define the mapping $G : \mathbb{R}^N \to \mathbb{R}^N$ by

$$G(x) = \left(\left\langle F\left(\sum_{i=1}^{n} x_i v_i\right), v_1\right\rangle, \ldots, \left\langle F\left(\sum_{i=1}^{n} x_i v_i\right), v_N\right\rangle\right) \quad \text{for all } x = (x_1, \ldots, x_N) \in \mathbb{R}^N.$$

Due to (2.15), the standard scalar product $G(x) \cdot x$ in $\mathbb{R}^N$ fulfills

$$G(x) \cdot x = \left\langle F\left(\sum_{i=1}^{N} x_i v_i\right), \sum_{i=1}^{N} x_i v_i \right\rangle \geq 0$$

for all $x = (x_1, \ldots, x_N) \in \mathbb{R}^N$ with

$$|x|_1 := \left\| \sum_{i=1}^{N} x_i v_i \right\|_V = R.$$

We note that $|\cdot|_1$ is a norm on $\mathbb{R}^N$ and consider the corresponding closed ball

$$B_R = \left\{ x \in \mathbb{R}^N : |x|_1 \leq R \right\}.$$

Our assumption in arguing by contradiction yields that the map $g : B_R \to B_R$ given by

$$g(x) = -\frac{R}{|G(x)|_1} G(x) \quad \text{for all } x \in B_R$$

is well defined and continuous. Consequently, the Brouwer Fixed Point Theorem, see Theorem 2.10, ensures the existence of some $x_0 \in B_R$ satisfying $g(x_0) = x_0$. Since

$$|x_0|_1 = |g(x_0)|_1 = R,$$

we have $x_0 \neq 0$. On the other hand, denoting by $|\cdot|_2$ the Euclidean norm on $\mathbb{R}^N$, there exists a constant $c > 0$ such that $|x|_1 \leq c|x|_2$ whenever $x \in \mathbb{R}^N$. This leads to a contradiction

$$0 < R^2 = |x_0|_1^2 \leq c^2|x_0|_2^2 = c^2 x_0 \cdot x_0 = c^2 x_0 \cdot g(x_0) = -\frac{Rc^2}{|G(x_0)|_1} G(x_0) \cdot x_0 \leq 0,$$

which proves the result. $\qquad\square$

Theorem 2.11 (Finite dimensional existence theorem) *Assume that $a \in \mathbb{R}^N$ is fixed and that $f : \mathbb{R}^N \to \mathbb{R}^N$ is coercive, continuous and monotone. Then the set K of solutions to the equation*

$$f(x) = a$$

is non-empty, closed and convex. The set K is a singleton in case f is strictly monotone.

Remark 2.12 In a finite dimensional setting for the existence of solutions monotonicity is not required. We would like to mention that in a finite dimensional setting surjectivity and monotonicity imply coercivity of a mapping. Precisely speaking, assume that $f : \mathbb{R}^N \to \mathbb{R}^N$ is surjective (i.e. $f\left(\mathbb{R}^N\right) = \mathbb{R}^N$) and monotone. Then f is weakly coercive, i.e.

$$\lim_{|x| \to +\infty} |f(x)| = +\infty. \tag{2.16}$$

Indeed, supposing to the contrary, there is a sequence $\{x_n\}_{n \geq 1} \subset \mathbb{R}^N$, $x_n \neq 0$, such that for some $M > 0$

$$|x_n| \to +\infty \text{ and } |f(x_n)| \leq M.$$

Define for $n \in \mathbb{N}$

$$w_n = \frac{x_n}{|x_n|}.$$

Since the sequence $\{w_n\}_{n \geq 1}$ is bounded, we can take a convergent subsequence which we denote by $\{w_n\}_{n \geq 1}$. Let its limit be w_0. Since $f\left(\mathbb{R}^N\right) = \mathbb{R}^N$ we see that there is x_0 such that

$$f(x_0) = (M+1) w_0.$$

Note that

$$\frac{x_n - x_0}{|x_n|} \to w_0 \text{ as } n \to +\infty \text{ and } \limsup_{n \to +\infty} |f(x_n)| \le M.$$

The monotonicity of f implies that

$$\left(f(x_n), \frac{x_n - x_0}{|x_n|} \right) \ge \left(f(x_0), \frac{x_n - x_0}{|x_n|} \right)$$

Taking lim sup to both sides of the above inequality, we reach a contradiction.

Note that the function $f : \mathbb{R} \to \mathbb{R}$ defined by

$$f(x) = \begin{cases} x \sin x, & x \ge 0 \\ 0, & x < 0 \end{cases}$$

does not satisfy condition (2.16), but it is surjective. We conclude this remark with some additional comment. Define $g : \mathbb{R} \to \mathbb{R}$ by

$$g(x) = \begin{cases} x, & x < 0, \\ x + 1, & x \ge 0. \end{cases}$$

Consider the operator $f : \mathbb{R}^2 \to \mathbb{R}^2$ given by

$$f(x, y) = (y + g(x), -x).$$

Then f is monotone and surjective but it is not continuous.

In a finite dimensional setting we have the following consequence of Lemma 2.12 which provides existence of solutions for example for difference equations, discrete equations and related problems.

Theorem 2.12 (Finite dimensional existence) *Let E be a finite dimensional real Banach space and let $A : E \to E^*$ be a continuous mapping. Assume that there exists a function $r : [0, +\infty) \to \mathbb{R}$ such that $\lim_{t \to +\infty} r(t) = +\infty$ and that the inequality*

$$(A(x), x) \ge r(\|x\|) \|x\| \tag{2.17}$$

holds for all $x \in E$. Then for each $g \in E^*$ the equation $A(x) = g$ has at least one solution.

Here is the result connecting the continuity notions for monotone operators as well as to the idea of the maximal monotonicity.

Proof It suffices to assume that $g = 0$. Relation (2.17) implies that the operator A is coercive. Hence there is some $R > 0$ such that (2.15) is satisfied with $A = F$. Thus the result follows from Lemma 2.12. □

Lemma 2.13 *(Minty Lemma) Assume that the operator $A : E \to E^*$ is monotone. Then the following assertions are equivalent:*

(i) *the operator A is radially continuous;*
(ii) *the relation $\langle f - A(v), u_0 - v \rangle \geq 0$ for all $v \in E$ implies that $A(u_0) = f$;*
(iii) *the relations $u_n \rightharpoonup u_0$ in E, $A(u_n) \rightharpoonup f$ in E^* and*

$$\limsup_{n \to +\infty} \langle A(u_n), u_n \rangle \leq \langle f, u_0 \rangle$$

   imply that $A(u_0) = f$;
(iv) *the operator A is demicontinuous.*

At this point we are able to obtain the existence result.

Theorem 2.13 (Browder-Minty Theorem) *Assume that the operator $A : E \to E^*$ is radially continuous, coercive and monotone. Then for any $f \in E^*$ the set K of solutions to the equation*

$$A(u) = f$$

is non-empty, bounded, sequentially weakly closed, and convex.

Proof In view of Lemmas 2.9 and 2.10, it remains to examine the solvability of the equation $A(u) = f$.

Let us fix $n \in \mathbb{N}$ and the space E_n from Remark 2.11. By f_n we denote the restriction of the functional f to the space E_n. Similarly by A_n we denote the restriction of A to the space E_n. Then there holds

$$A_n : E_n \to E_n^*.$$

Note that since A is coercive and monotone, so is A_n. Moreover, since by Lemma 2.13 the operator A is demicontinuous, it follows that A_n is continuous. Thus by Theorem 2.11 the equation

$$A_n(u) = f_n$$

has at least one solution u_n. By the definition of A_n this means that

$$\langle A(u_n), h_k \rangle = \langle f, h_k \rangle \text{ for } k = 1, 2, .., n. \tag{2.18}$$

Consequently, the sequence $\{u_n\}_{n \geq 1}$ -called the sequence of the Galerkin approximations - satisfies

$$\langle A(u_n), u_n \rangle = \langle f_n, u_n \rangle = \langle f, u_n \rangle \leq \|f\|_* \|u_n\|. \tag{2.19}$$

Since the operator A is coercive we observe that the sequence $\{u_n\}_{n \geq 1}$ is bounded. Thus there is some $R > 0$ such that

$$\|u_n\| \leq R \text{ for } n \in \mathbb{N}.$$

From (2.19) we infer that

$$\langle A(u_n), u_n \rangle = \langle f, u_n \rangle \leq \|f\|_* R.$$

Taking into account that $\|u_n\| \leq R$ and

$$\langle A(u_n), u_n \rangle \leq \|f\|_* R$$

we have by Lemma 2.11 that $\|A(u_n)\|_* \leq M_1$ for some fixed $M_1 > 0$ and for all $n \in \mathbb{N}$. Next, we see that from (2.18) it follows

$$\lim_{n \to +\infty} \langle A(u_n), h \rangle = \langle f, h \rangle \text{ for each } h \in \bigcup_n E_n.$$

This amounts to saying that $A\{u_n\} \rightharpoonup f$. Since $\{u_n\}_{n \geq 1}$ is bounded, there is a subsequence $\{u_{n_k}\}_{k \geq 1}$ weakly convergent to some $u_0 \in E$. Again from (2.18) we see for all $k \in \mathbb{N}$ that

$$\langle A(u_{n_k}), u_{n_k} \rangle = \langle f, u_{n_k} \rangle.$$

Passing to the limit, we get

$$\lim_{k \to +\infty} \langle A(u_{n_k}), u_{n_k} \rangle = \langle f, u_0 \rangle.$$

Hence Lemma 2.13 (iii) entails that $A(u_0) = f$ completing the proof. $\qquad \square$

Remark 2.13 Concerning the proof of Theorem 2.13 we have the following observations:

(i) The proof of Theorem 2.13 is constructive since the approximations represent the basis for many numerical methods.
(ii) The entire approximating sequence from the above theorem need not be even weakly convergent unless the uniqueness is guaranteed.
(iii) The assumption of reflexivity cannot be easily omitted due to the required weak compactness of the closed ball. The assumption of separability is not essential and can be overcome.
(iv) When A is strictly monotone, the entire sequence $\{u_n\}_{n \geq 1}$ constructed in the above proof is weakly convergent.
(v) When A satisfies condition (S) or related, only a certain subsequence $\{u_{n_k}\}_{k \geq 1}$ converges in the norm unless strict monotonicity is assumed.

2.8   Pseudomonotone Operators

The monotonicity condition is not required in the proof of the Minty-Browder Theorem. It can be replaced by various other assumptions.

Definition 2.10 *(Condition (M))* Consider an operator $A : E \to E^*$. If the relations

$$u_n \rightharpoonup u_0 \text{ in } E \text{ and } A(u_n) \rightharpoonup f \text{ in } E^*$$

and

$$\limsup_{n \to +\infty} \langle A(u_n), u_n \rangle \leq \langle f, u_0 \rangle$$

imply $A(u_0) = f$, then A is said to satisfy condition (M) or that A is of type (M).

If $A : E \to E^*$ is a demicontinuous operator of type $(S)_+$ and $C : E \to E^*$ is a compact operator, then $A + C$ is of type (M).

Corollary 2.2 *Assume that $A : E \to E^*$ is bounded, coercive and satisfies condition (M). Then for any $f \in E^*$ the set K of solutions to*

$$A(u) = f$$

is non-empty and bounded.

In connection to the above corollary a related notion arises.

Definition 2.11 *(Pseudomonotone operator)* We say that the operator $A : E \to E^*$ is pseudomonotone when the following implication holds:
   if

$$u_n \rightharpoonup u_0 \text{ in } E$$

and

$$\limsup_{n \to +\infty} \langle A(u_n), u_n - u_0 \rangle \leq 0,$$

then

$$\liminf_{n \to +\infty} \langle A(u_n), u_n - v \rangle \geq \langle A(u_0), u_0 - v \rangle \text{ for all } v \in E.$$

A strongly continuous operator is pseudomonotone. The sum of two pseudomonotone operators is a pseudomonotone operator. This means that strongly continuous perturbations of continuous monotone operators are pseudomonotone, but not necessarily monotone. If an operator $A : E \to E^*$ is monotone and hemicontinuous, then it is pseudomonotone. On the other hand, we have what follows:

Lemma 2.14 *If an operator $A : E \to E^*$ is pseudomonotone and bounded, then it is demicontinuous.*

The proof can be found in [73], page 28. If an operator is pseudomonotone, then it satisfies condition (M). Therefore we have the following version of the Browder-Minty Theorem involving pseudomonotone operators:

Theorem 2.14 (Surjectivity theorem for pseudomonotone operators) *Assume that $A : E \to E^*$ is pseudomonotone, bounded and coercive. Then for any fixed $g \in E^*$ the set K of solutions to*

$$A(u) = g$$

is non-empty.

2.9 An Application to Solvability Through Pseudomonotone Operators

In this section we apply Theorem 2.14 to prove the existence of a solution to the following quasilinear Dirichlet problem

$$\begin{cases} -\mathrm{div}(a(x)|\nabla u|^{p-2}\nabla u + b(x)|\nabla u|^{q-2}\nabla u) = f(x, u, \nabla u) & \text{in } \Omega \\ u = 0 & \text{on } \partial\Omega \end{cases} \tag{2.20}$$

on a bounded domain $\Omega \subset \mathbb{R}^N$ with $N \geq 1$ and a Lipschitz boundary $\partial\Omega$. In the principal part of the equation in (2.20) we have two nonnegative weights $a \in L^1(\Omega)$ and $b \in L^1(\Omega)$ entering the operator in divergence form denoted $-\Delta_{p,q}^{a,b}$ in Section 2.4, where $1 < q < p < +\infty$. According to (2.13) we can express problem (2.20) as

$$\begin{cases} -\Delta_{p,q}^{a,b}u = f(x, u, \nabla u) & \text{in } \Omega \\ u = 0 & \text{on } \partial\Omega. \end{cases}$$

As before, to simplify the presentation, for any real number $r > 1$ we denote $r' := r/(r - 1)$ (the Hölder conjugate of r).

Lemma 2.15 *Assume that hypothesis (H)—see (1.1)- and condition (1.5) hold. In addition, assume that $f : \Omega \times \mathbb{R} \times \mathbb{R}^N \to \mathbb{R}$ is a Carathéodory function (i.e., $f(\cdot, t, \xi)$ is measurable on Ω for each $(t, \xi) \in \mathbb{R} \times \mathbb{R}^N$ and $f(x, \cdot, \cdot)$ is continuous on $\mathbb{R} \times \mathbb{R}^N$ for a.e. $x \in \Omega$) satisfying:*

$$|f(x, t, \xi)| \leq \sigma(x) + c_1|t|^\alpha + c_2|\xi|^\beta \tag{2.21}$$

for a.e $x \in \Omega$, all $(t, \xi) \in \mathbb{R} \times \mathbb{R}^N$, with $\sigma \in L^{\gamma'}(\Omega)$ for some $\gamma \in (1, p_s^*)$ and constants

$$c_1 > 0, \quad c_2 > 0, \quad \alpha \in [0, p_s^* - 1), \quad \beta \in [0, \frac{p_s}{(p_s^*)'}).$$

Set

$$\theta := \min\left\{\gamma', \frac{p_s^*}{\alpha}, \frac{p_s}{\beta}\right\}. \tag{2.22}$$

Then the Niemytskij operator

$$\mathcal{N}_f : L^{p_s^*}(\Omega) \times (L^{p_s}(\Omega))^N \to L^\theta(\Omega)$$

associated with the function f, that is,

$$\mathcal{N}_f(v, z) = f(\cdot, v(\cdot), z(\cdot)) \text{ for all } v, z \in L^{p_s^*}(\Omega) \times (L^{p_s}(\Omega))^N \tag{2.23}$$

is well defined, continuous and bounded.

Proof From (2.22) we see that $\theta > 1$. Indeed, $\gamma' > 1$ (note that $\gamma > 1$),

$$\frac{p_s^*}{\alpha} > \frac{p_s^*}{p_s^* - 1} = (p_s^*)' > 1, \quad \frac{p_s}{\beta} > (p_s^*)' > 1$$

(note $p_s^* > p > 1$).

We observe from (2.21) that

$$|f(x, t, \xi)| \leq \tilde{\sigma}(x) + c_1 |t|^{\frac{p_s^*}{\theta}} + c_2 |\xi|^{\frac{p_s}{\theta}} \tag{2.24}$$

for a.e $x \in \Omega$, all $(t, \xi) \in \mathbb{R} \times \mathbb{R}^N$, with $\tilde{\sigma} \in L^\theta(\Omega)$. In order to show this, we note from (2.22) that

$$\alpha \leq \frac{p_s^*}{\theta} \text{ and } \beta \leq \frac{p_s}{\theta},$$

so (2.22) is deduced from (2.21) setting $\tilde{\sigma}(x) = \sigma(x) + c_1 + c_2$ for a.e. $x \in \Omega$ that gives $\tilde{\sigma} \in L^{\gamma'}(\Omega) \subset L^\theta(\Omega)$. The Krasnosel'skii Theorem (see Theorem 1.1) provides that the Niemytskij operator $\mathcal{N}_f$ defined by (2.23) fulfills the required properties. $\qquad\square$

Let us introduce the map $N_f : W^{1,p}(a, \Omega) \to W^{1,p}(a, \Omega)^*$ by

$$\langle N_f(u), v\rangle_{W^{1,p}(a,\Omega)} = \int_\Omega f(x, u(x), \nabla u(x))v(x)dx \text{ for all } u, v \in W_0^{1,p}(a, \Omega).$$

Using (1.4) we have

$$(u, \nabla u) \in L^{p_s^*}(\Omega) \times (L^{p_s}(\Omega))^N \text{ whenever } u \in W_0^{1,p}(a, \Omega),$$

whence (2.23) implies that

$$\langle N_f(u), v \rangle_{W^{1,p}(a,\Omega)} = \langle \mathcal{N}_f(u, \nabla u), v \rangle_{L^\theta(\Omega)} \tag{2.25}$$

for all $u, v \in W_0^{1,p}(a, \Omega)$.

Now we define the map

$$A : W_0^{1,p}(a, \Omega) \to W_0^{1,p}(a, \Omega)^*$$

as

$$A = -\Delta_{p,q}^{a,b} - N_f. \tag{2.26}$$

Proposition 2.3 *Assume that hypothesis (H) and conditions (1.1) and (2.21) hold. Then the map $A : W_0^{1,p}(a, \Omega) \to W_0^{1,p}(a, \Omega)^*$ introduced in (2.26) is a pseudomonotone operator.*

Proof According to Definition 2.11, suppose that $u_n \rightharpoonup u$ in $W_0^{1,p}(a, \Omega)$ and

$$\limsup_{n \to \infty} \langle A(u_n), u_n - u \rangle \leq 0. \tag{2.27}$$

By (2.25) we have that

$$|\langle N_f(u_n), u_n - u \rangle| \leq \|\mathcal{N}_f(u_n, \nabla u_n)\|_{L^\theta(\Omega)} \|u_n - u\|_{L^{\theta'}(\Omega)}. \tag{2.28}$$

Lemma 2.15 ensures that $\{\mathcal{N}_f(u_n, \nabla u_n)\}$ is bounded in $L^\theta(\Omega)$. On the other hand, it follows from (2.22) that

$$\theta' = \max\left\{ \gamma, \left(\frac{p_s^*}{\alpha}\right)', \left(\frac{p_s}{\beta}\right)' \right\}. \tag{2.29}$$

The assumptions in Lemma 2.15 render that $\gamma < p_s^*$,

$$\left(\frac{p_s^*}{\alpha}\right)' = \frac{\frac{p_s^*}{\alpha}}{\frac{p_s^*}{\alpha} - 1} = \frac{p_s^*}{p_s^* - \alpha} < p_s^*,$$

$$\left(\frac{p_s}{\beta}\right)' = \frac{\frac{p_s}{\beta}}{\frac{p_s}{\beta} - 1} = \frac{p_s}{p_s - \beta} < \frac{p_s}{p_s - \frac{p_s}{(p_s^*)'}} = \frac{(p_s^*)'}{(p_s^*)' - 1} = p_s^*.$$

Hence (2.29) entails $1 < \theta' < p_s^*$. Through the Rellich-Kondrachov compact embedding theorem we conclude that the embedding $W^{1,p_s}(\Omega) \hookrightarrow L^{\theta'}(\Omega)$ is compact, so $u_n \to u$ in $L^{\theta'}(\Omega)$. Then (2.27) and (2.28) produce (2.14). At this stage, Proposition 2.2 provides the strong convergence $u_n \to u$ in $W_0^{1,p}(a, \Omega)$. The continuity of the maps $\Delta_{p,q}^{a,b}$ and N_f from $W_0^{1,p}(a, \Omega)$ to $W_0^{1,p}(a, \Omega)^*$ implies that

$$A(u_n) \to A(u) \text{ in } W_0^{1,p}(a, \Omega)^*.$$

Taking into account (2.26), we infer that

$$\liminf_{n \to +\infty} \langle A(u_n), u_n - v \rangle \geq \langle A(u_0), u_0 - v \rangle \text{ for all } v \in W_0^{1,p}(a, \Omega),$$

which completes the proof. $\square$

Our existence result for problem (2.20) reads as follows.

Theorem 2.15 *Assume that the conditions in Proposition 2.3 are fulfilled together with*

$$f(x, t, \xi)t \leq \rho(x) + d_1|t|^p + d_2 a(x)|\xi|^p \tag{2.30}$$

for a.e $x \in \Omega$ and all $(t, \xi) \in \mathbb{R} \times \mathbb{R}^N$, with $\rho \in L^1(\Omega)$ and with constants $d_1 > 0$ and $d_2 > 0$ satisfying

$$(\lambda_1^a)^{-1} d_1 + d_2 < 1, \tag{2.31}$$

where λ_1^a stands for the first eigenvalue of $-\Delta_p^a$ on $W_0^{1,p}(a, \Omega)$. Then there exists a weak solution $u \in W_0^{1,p}(a, \Omega)$ to problem (2.20) in the sense that

$$\int_\Omega (a(x)|\nabla u|^{p-2}\nabla u + b(x)|\nabla u|^{q-2}\nabla u)\nabla v\,dx = \int_\Omega f(x, u, \nabla u)v\,dx \tag{2.32}$$

for all $v \in W_0^{1,p}(a, \Omega)$.

Proof Since the operators $\Delta_{p,q}^{a,b}$ and $N_f : W_0^{1,p}(a, \Omega) \to W_0^{1,p}(a, \Omega)^*$ are bounded, it follows that the operator $A : W_0^{1,p}(a, \Omega) \to W_0^{1,p}(a, \Omega)^*$ introduced in (2.26) is bounded. Proposition 2.3 ensures that the operator A is pseudomonotone.

Next we show that the operator $A : W_0^{1,p}(a, \Omega) \to W_0^{1,p}(a, \Omega)^*$ is coercive, that is,

$$\lim_{\|u\|_{W_0^{1,p}(a,\Omega)} \to +\infty} \frac{\langle A(u), u \rangle}{\|u\|_{W_0^{1,p}(a,\Omega)}} = +\infty. \tag{2.33}$$

From (2.26), (2.13) and (2.30) we derive the estimate

$$\langle A(u), u \rangle = \langle -\Delta_{p,q}^{a,b}(u), u \rangle - \int_\Omega f(x, u, \nabla u)u\,dx$$

$$\geq \|u\|_{W_0^{1,p}(a,\Omega)}^p - \|\rho\|_{L^1(\Omega)} - d_1\|u\|_{L^p(\Omega)}^p - d_2\|u\|_{W_0^{1,p}(a,\Omega)}^p$$

$$\geq (1 - d_1(\lambda_1^a)^{-1} - d_2)\|u\|_{W_0^{1,p}(a,\Omega)}^p - \|\rho\|_{L^1(\Omega)}.$$

Then assumption (2.31) leads to claim (2.33) because $p > 1$.

Therefore the operator $A : W_0^{1,p}(a, \Omega) \to W_0^{1,p}(a, \Omega)^*$ is bounded, pseudomonotone and coercive. This enables us to apply Theorem 2.14 ensuring that A is surjective. In particular, there exists $u \in W_0^{1,p}(\Omega)$ such that $A(u) = 0$, which is just (2.32), thus completing the proof. $\qquad\square$

We end this section with an example taken from [56]. We consider the following Dirichlet problem

$$
\begin{cases}
-\mathrm{div}(|x|^{\frac{1}{3}}|\nabla u|\nabla u + (1 - |x|^2)\nabla u) = g(x, u) - ku|\nabla u|^{\frac{2}{3}} & \text{in } B \\[2mm]
u = 0 & \text{on } \partial B
\end{cases}
\tag{2.34}
$$

on the unit ball

$$
B = \{x \in \mathbb{R}^3 : |x| < 1\},
$$

with a constant $k > 0$ and a Carathéodory function $g : B \times \mathbb{R} \to \mathbb{R}$ satisfying

$$
|g(x, t)| \leq c_0 t^2 + c \text{ for a.e } x \in B, \text{ all } t \in \mathbb{R},
$$

with constants $c_0 \geq 0$ and $c \geq 0$ provided $c_0 + c < \lambda_1^a$, where λ_1^a denotes the first eigenvalue of $-\Delta_3^a$ on $W_0^{1,3}(a, B)$. We easily see that statement (2.34) fits problem (2.20) with $N = 3$, $p = 3$, $q = 2$, $\Omega = B$, and with

$$
a(x) = |x|^{\frac{1}{3}}, b(x) = 1 - |x|^2 \text{ and } f(x, t, \xi) = g(x, t) - kt|\xi|^{\frac{2}{3}}.
$$

Assumption (H) is verified because

$$
a^{-\frac{q}{p-q}} b^{\frac{p}{p-q}} = |x|^{-\frac{2}{3}}(1 - |x|^2)^3 \in L^1(B).
$$

Assumption (1.5) requires $a^{-s} = |x|^{-\frac{s}{3}} \in L^1(B)$ for some

$$
s \in \left(\max\left\{ \frac{N}{p}, \frac{1}{p-1} \right\}, +\infty \right) = (1, +\infty),
$$

which amounts to choosing $1 < s < 9$, for instance $s = 2$. With this choice, we have

$$
p_s = ps/(s + 1) = 2.
$$

Thus $N = 3 > p_s = 2$, which implies

$$
p_s^* = \frac{N p_s}{N - p_s} = 6, \ (p_s^*)' = \frac{6}{5} \text{ and } \frac{p_s}{(p_s^*)'} = \frac{5}{3}.
$$

It holds

$$|f(x,t,\xi)| \leq c_0 t^2 + c + k|t||\xi|^{\frac{2}{3}} \leq (c_0 + \frac{k}{2})t^2 + c + \frac{k}{2}|\xi|^{\frac{4}{3}}$$

for a.e $x \in \Omega$ and all $(t,\xi) \in \mathbb{R} \times \mathbb{R}^3$. Consequently, assumption (2.24) is satisfied with $\sigma(x) = c$, $c_1 = c_0 + \frac{k}{2}$, $c_2 = \frac{k}{2}$, $\alpha = 2$ and $\beta = \frac{4}{3}$. Moreover, we have

$$f(x,t,\xi)t = g(x,t)t - kt^2|\xi|^{\frac{2}{3}} \leq (c_0 t^2 + c)|t| \leq (c_0 + c)|t|^3 + c$$

for a.e $x \in \Omega$ and all $(t,\xi) \in \mathbb{R} \times \mathbb{R}^3$, so requirement (2.30) is also verified. As a result, Theorem 2.15 ensures that problem (2.34) possesses at least a weak solution in the sense of (2.32).

For a more comprehensive treatment of these methods, we refer to [31, 49, 57, 62, 75]. We base ourselves here on these sources mentioned.

Chapter 3
Generalized Solutions for Non-potential Problems

Abstract The objective of this chapter is to develop an approach for proving the existence of solutions to problems where the monotonicity property of the driving operator is absent. We focus on nonlinear boundary value problems driven by continuous, coercive operators that lack sufficient monotonicity or pseudo-monotonicity to permit the application of the Minty–Browder theorem, in either version discussed in the preceding chapter. Assuming that the given operator is continuous on finite-dimensional subspaces and coercive, we immediately obtain the solvability of each Galerkin-type approximation, along with the boundedness of the sequence of approximate solutions. Although we can extract a weakly convergent subsequence, we cannot assert the solvability of the original nonlinear problem in the usual (weak) sense. The concept of a solution, referred to as a generalized solution, is introduced in this chapter. We begin by describing competing operators and then consider a model boundary value problem driven by the competing (p, q)-Laplacian with Dirichlet boundary conditions, examining the convergence of the associated Galerkin scheme. Next, we introduce the notion of an abstract generalized solution, whose existence is guaranteed by the coercivity and continuity of the operator. In addition to the model problem, which also involves an unbounded weight, we present applications to competing fractional operators and mixed local–nonlocal competing problems.

Keywords Abstract generalized solution · Competing (p, q)-Laplacian · Convection · Dirichlet problem · Galerkin basis · Generalized solution · Non-potential operator · Strong generalized solution · Truncation operator · Unbounded weight

3.1 On the Competing (p, q)-Laplacian

A careful examination of the proof of Theorem 2.13 (the Browder-Minty Theorem) reveals that the monotonicity of the operator A is used in the following steps:

© The Author(s), under exclusive license to Springer Nature Switzerland AG 2026

M. Galewski and D. Motreanu, *Competing Operators and Their Applications to Boundary Value Problems*, SpringerBriefs in Mathematics,
https://doi.org/10.1007/978-3-032-15445-3_3

(a) the continuity of A_n being a consequence of the fact that a radially continuous monotone operator is demicontinuous;
(b) local boundedness of A which is a consequence of the monotonicity (see Proposition 2.1);
(c) assertion (iii) of Lemma 2.13 related to condition (M).

We can deal with the continuity by replacing radial continuity with just a continuity assumption, in which case the property mentioned as (a) still holds. This is just a slight strenghtening of the assumptions and does not influence much the applicability. As far as (b) is concerned, we can assume that the operator is bounded. This property typically follows from the growth assumptions and again does not restrict the range of nonlinear terms to be considered. Note that differential operators are usually bounded. The most demanding is the condition written above as (c) and by ignoring it we obtain our new solvability notion. There is also no suitable replacement of it contrary to cases (a) and (b). A prototype of operators which lack monotonicity in the principal part of the equation is $-\Delta_p + \Delta_q$ on $W_0^{1,p}(\Omega)$, with $1 < q < p < +\infty$ and a bounded domain $\Omega \subset \mathbb{R}^N$. Specifically, we see that

$$- \Delta_p + \Delta_q : W_0^{1,p}(\Omega) \to W_0^{-1,p'}(\Omega)$$

and that $-\Delta_p + \Delta_q$ is continuous due to the standard properties of the negative p-Laplacian, see Theorem 2.8, and due to the continuous embedding of $W_0^{1,p}(\Omega)$ into $W_0^{1,q}(\Omega)$. The latter result provides also that the operator $-\Delta_p + \Delta_q$ is coercive. Indeed, for any $u \in W_0^{1,p}(\Omega)$ we have for some constant $c > 0$ (which follows from the embedding):

$$\langle -\Delta_p u + \Delta_q u, u \rangle = \|\nabla u\|_p^p - \|\nabla u\|_q^q \geq \|\nabla u\|_p^p - c \|\nabla u\|_p^q .$$

However, there are some differences with the negative p-Laplacian and with the (p, q)-Laplacian. Namely, we see that taking a nonzero element $u_0 \in W_0^{1,p}(\Omega)$ and a real number $\lambda > 0$ the expression

$$\langle -\Delta_p (\lambda u_0) + \Delta_q (\lambda u_0) , \lambda u_0 \rangle = \lambda^p \|\nabla u_0\|_p^p - \lambda^q \|\nabla u_0\|_q^q$$

is not of constant sign, being positive for λ sufficiently large and negative otherwise, thus the ellipticity is lost. For this reason we call the operator $-\Delta_p + \Delta_q$ the competing (p, q) -Laplacian. We also note that this operator is in divergence form

$$- \Delta_p u + \Delta_q u = -div(a(|\nabla u|)\nabla u) \text{ with } a(t) = t^{p-2} - t^{q-2} \text{ for all } t > 0.$$

A minimal condition of ellipticity for an operator in divergence form $-div(a(|\nabla u|\nabla u))$ is to have, among other things, $a(t) > 0$ for all $t > 0$, which is obviously not satisfied in the case of

$$a(t) = t^{p-2} - t^{q-2}.$$

Nevertheless, the problems on which we focus are nonvariational due to the fact that they involve convection terms $f(x, u, \nabla u)$ where the presence of the gradient prevents the construction of a potential. A systematic study of nonvariational methods applied to elliptic problems can be found in [50]. The difficulty in problems that we consider is to deal simultaneously with competing operators and convection terms. We overcome the difficulty by resolving finite dimensional approximated problems and then passing to the limit in an appropriate sense.

The lack of any monotonicity property for the driving operator $-\Delta_p + \Delta_q$ prevents to apply the surjectivity result for pseudomonotone operators in Theorem 2.14. There is a striking difference between the operators $-\Delta_p - \Delta_q$ and $-\Delta_p + \Delta_q$.

The operator $-\Delta_p - \Delta_q$ is strictly monotone and continuous as can be verified easily with Theorem 2.8, so it is pseudomonotone, whereas this result fails for $-\Delta_p + \Delta_q$ as mentioned above. In this connection, let us note that even the linear continuous operator Δ on $H_0^1(\Omega)$ is not pseudomonotone. Assuming to the contrary, if $u_n \rightharpoonup u$ in $H_0^1(\Omega)$ with

$$\limsup_{n \to \infty} \langle \Delta u_n, u_n - u \rangle \leq 0,$$

the pseudomonotonicity of Δ would mean that

$$\liminf_{n \to \infty} \langle \Delta u_n, u_n - v \rangle \geq \langle \Delta u, u - v \rangle$$

whenever $v \in H_0^1(\Omega)$. Thus in particular we would have $\liminf_{n \to \infty} \langle \Delta u_n, u_n \rangle \geq \langle \Delta u, u \rangle$. The inequality

$$\limsup_{n \to \infty} \langle \Delta u_n, u_n - u \rangle \leq 0$$

is always true being equivalent to

$$\liminf_{n \to \infty} \| \nabla u_n \|_{L^2(\Omega)}^2 \geq \| \nabla u \|_{L^2(\Omega)}^2$$

that holds thanks to the sequential weak lower semicontinuity of the norm. Besides,

$$\liminf_{n \to \infty} \langle \Delta u_n, u_n \rangle \geq \langle \Delta u, u \rangle$$

is equivalent to

$$\limsup_{n \to \infty} \| \nabla u_n \|_{L^2(\Omega)}^2 \leq \| \nabla u \|_{L^2(\Omega)}^2,$$

which entails the strong convergence $u_n \to u$ since the space $H_0^1(\Omega)$ is uniformly convex. Thus, we would reach the conclusion that every weakly convergent sequence is strongly convergent. The contradiction obtained proves the claim.

The deficit of ellipticity, monotonicity and variational structure is resolved by seeking the solution as a limit of finite dimensional approximations that we call a generalized solution. It is instructive to observe that the same procedure applied to a problem driven by the ordinary (p, q)-Laplacian $-\Delta_p - \Delta_q$ leads to the existence of a weak solution. Applications are given in [24–26, 32, 45, 58]. Related results can be found in [2, 6, 53].

3.2   A Model Problem

Before the introduction of a general existence tool concerning the solvability of problems that lack monotonicity in the sense of the previous section, we deal with some model problems by highlighting their main properties. Thus, our goal is to study the following problem with homogeneous Dirichlet boundary condition and an unbounded term $g(u)$ in the differential operator of competing type

$$
\begin{cases}
- \operatorname{div}\left(g(u\,(x))|\nabla u\,(x)\,|^{p-2}\nabla u\,(x)\right) + \operatorname{div}\left(|\nabla u\,(x)\,|^{q-2}\nabla u\,(x)\right) \\
= f(x, u\,(x)\,, \nabla u\,(x)) & \text{in } \Omega, \\
u\,(x) = 0 & \text{on } \partial\Omega,
\end{cases}
\tag{3.1}
$$

with $1 < q < p < +\infty$. Problem (3.1) is considered on a bounded domain $\Omega \subset \mathbb{R}^N$ with Lipschitz boundary $\partial\Omega$. We can resolve (3.1) in the case $N < p$ in the presence of an unbounded weight $g\,(u)$. In the absence of a weight, we are able to remove this assumption. We will address such a problem in the sequel to illustrate the abstract scheme, which we will introduce later.

The approach employed here follows [15]. Since we assume that $N < p$, there is a continuous embedding of $W_0^{1,p}(\Omega)$ into $C\left(\overline{\Omega}\right)$, so there is a positive constant C_S such that the following Sobolev inequality holds

$$
\max_{x \in \overline{\Omega}} |u\,(x)| \leq C_S \|\nabla u\|_{L^p(\Omega)}
\tag{3.2}
$$

for all $u \in W_0^{1,p}(\Omega)$.

The weight g employed in (3.1) satisfies the assumption:

(H1) $g : \mathbb{R} \to [a_0, +\infty)$ *is a continuous function with* $a_0 > 0$.

In (3.1) the convection term is expressed through a Carathéodory function

$$
f : \Omega \times \mathbb{R} \times \mathbb{R}^N \to \mathbb{R}
$$

which is subject to the growth conditions:

(H2)   *There exist a nonnegative function $\sigma \in L^{r_1}(\Omega)$ and constants $b \geq 0$ and $c \geq 0$ such that*

$$|f(x, s, \xi)| \leq \sigma(x) + b|s|^{r_2} + c|\xi|^{p-1} \quad \text{for a.e.} x \in \Omega, \text{all } s \in \mathbb{R}, \xi \in \mathbb{R}^N,$$

with constants $r_1, r_2 \geq 1$.

(H3)   *There exist constants $c_0 < a_0, c_1 > 0$ and $\alpha \in [1, p)$ such that*

$$f(x, s, \xi)s \leq c_0|\xi|^p + c_1\left(|s|^\alpha + 1\right) \text{for a.e.} x \in \Omega, \text{ all } s \in \mathbb{R}, \ \xi \in \mathbb{R}^N.$$

Example 3.1 An example of a function $f : \Omega \times \mathbb{R} \times \mathbb{R}^N \to \mathbb{R}$ satisfying conditions **(H2)–(H3)** is

$$f(x, s, \xi) = |s|^{\alpha-2}s + \frac{s}{1 + s^2}\left(|\xi|^{p-1} + h(x)\right) \text{ for all } (x, s, \xi) \in \Omega \times \mathbb{R} \times \mathbb{R}^N,$$

with a constant $\alpha \in [1, p)$ and some $h \in L^\infty(\Omega)$. Further on we will address also another version of assumption **(H3)** in which we will handle the case when $\alpha = p$ which will imply some relation between c_1 and the first eigenvalue λ_1 of $-\Delta_p$.

The operator appearing in the left hand side of (3.1) with $g \equiv 1$ is the competing operator (p, q)-Laplacian which is non-monotone. An additional difficulty lies in the presence of the unbounded weight. We will handle it by means of the truncation

$$g_R(t) = \begin{cases} g(t) & \text{if } t \in [-R, R] \\ g(R) & \text{if } t > R \\ g(-R) & \text{if } t < -R \end{cases} \tag{3.3}$$

for to some constant $R > 0$. Corresponding to (3.1), we consider the auxiliary problem:

$$\begin{cases} -\operatorname{div}\left(g_R(u(x))|\nabla u(x)|^{p-2}\nabla u(x)\right) + \operatorname{div}\left(|\nabla u(x)|^{q-2}\nabla u(x)\right) \\ = f(x, u(x), \nabla u(x)) & \text{in } \Omega, \\ u(x) = 0 & \text{on } \partial\Omega. \end{cases} \tag{3.4}$$

We will show that (3.1) and (3.4) have bounded solutions with the same bounds. The solvability of these problems is treated by finding the limit of the Galerkin type approximations. Further on we will introduce some abstract tool leading to the existence of generalized solutions by dropping the monotonicity related assumption in a version of the Browder-Minty Theorem while retaining the continuity and the coercivity assumptions. This abstract tool will appear somehow parallel to the general scheme suggested by the application of the Minty-Browder Theorem.

We seek the solutions to problem (3.1) in $W_0^{1,p}(\Omega)$. The weak solution to (3.1), should it exist, is defined as follows

$$\int_\Omega g(u(x))|\nabla u(x)|^{p-2}\nabla u(x)\,\nabla v(x)\,dx - \int_\Omega |\nabla u(x)|^{q-2}\nabla u(x)\,\nabla v(x)\,dx$$

$$= \int_\Omega f(x,u(x),\nabla u(x))v(x)\,dx$$

for all $v \in W_0^{1,p}(\Omega)$. Such a solution however cannot be reached directly due to the mentioned lack of monotonicity whilst the variational methods are not applicable here. The operator

$$A : W_0^{1,p}(\Omega) \to W^{-1,p'}(\Omega)$$

given by

$$\langle A(u), v \rangle = \int_\Omega g(u(x))|\nabla u(x)|^{p-2}\nabla u(x)\,\nabla v(x)\,dx$$

$$- \int_\Omega |\nabla u(x)|^{q-2}\nabla u(x)\,\nabla v(x)\,dx - \int_\Omega f(x,u(x),\nabla u(x))v(x)\,dx$$

$$(3.5)$$

corresponds to problem (3.1). Observe that due to the presence of the weight, this operator is not bounded, however due to inequality (3.2) it is well posed.

3.3 Estimates for the Model Problem

We proceed with the following lemmas that will further lead to the well posedness and the coercivity of the operator A introduced in (3.5).

Lemma 3.1 *Under assumption **(H2)** there is a constant $C > 0$ such that*

$$\left| \int_\Omega f(x,u(x),\nabla u(x))v(x)\,dx \right|$$

$$\leq \int_\Omega |f(x,u(x),\nabla u(x))v(x)|\,dx$$

$$\leq C \left(\|\sigma\|_{L^{r_1}(\Omega)} + \|u\|_{L^{r_2}(\Omega)}^{r_2} + \|\nabla u\|_{L^p(\Omega)}^{p-1} \right) \|\nabla v\|_{L^p(\Omega)}$$

for all $u, v \in W_0^{1,p}(\Omega)$.

Proof Via assumption **(H2)** we get

$$\int_{\Omega} |f(x, u(x), \nabla u(x))v(x)|\, dx \leq \int_{\Omega} |\sigma(x)||v(x)|\, dx + b \int_{\Omega} |u(x)|^{r_2}|v(x)|\, dx$$

$$+ c \int_{\Omega} |\nabla u(x)|^{p-1}|v(x)|\, dx.$$

$$(3.6)$$

For the first term of right hand side of (3.6), the Sobolev inequality (3.2) and the Hölder inequality yield

$$\int_{\Omega} |\sigma(x)||v(x)|\, dx \leq \|v\|_{C(\overline{\Omega})} \int_{\Omega} |\sigma(x)|\, dx \leq C_S |\Omega|^{\frac{r_1-1}{r_1}} \|\sigma\|_{L^{r_1}(\Omega)} \|\nabla v\|_{L^p(\Omega)}.$$

Again by the Sobolev inequality (3.2) applied to the second summand we obtain

$$\int_{\Omega} |u(x)|^{r_2}|v(x)|\, dx \leq \|v\|_{C(\overline{\Omega})} \|u\|_{L^{r_2}(\Omega)}^{r_2} \leq C_S \|u\|_{L^{r_2}(\Omega)}^{r_2} \|\nabla v\|_{L^p(\Omega)}.$$

Finally, by the Hölder inequality and the variational expression of the first eigenvalue λ_1 of $-\Delta_p$ on $W_0^{1,p}(\Omega)$ we have

$$\int_{\Omega} |\nabla u(x)|^{p-1}|v(x)|\, dx \leq \|\nabla u\|_{L^p(\Omega)}^{p-1} \|v\|_{L^p(\Omega)} \leq \frac{1}{\lambda_1^{1/p}} \|\nabla u\|_{L^p(\Omega)}^{p-1} \|\nabla v\|_{L^p(\Omega)}.$$

Combining the above estimates we obtain the stated assertion. $\square$

Corollary 3.1 *Under assumption **(H2)** the Niemytskij operator*

$$N_f : W_0^{1,p}(\Omega) \to W^{-1,p'}(\Omega)$$

induced by the Carathéodory function

$$f : \Omega \times \mathbb{R} \times \mathbb{R}^N \to \mathbb{R},$$

namely,

$$N_f(w) = f(\cdot, w(\cdot), \nabla w(\cdot)), \quad \forall w \in W_0^{1,p}(\Omega),$$

is well defined, continuous and such that there exists a constant $C > 0$ for which it holds

$$\|N_f(u)\|_{W^{-1,p'}(\Omega)} \leq C \left(\|\sigma\|_{L^{r_1}} + \|u\|_{L^{r_2}}^{r_2} + \|\nabla u\|_{L^p}^{p-1} \right) \quad \textit{for all } u \in W_0^{1,p}(\Omega).$$

Now we proceed to some bounds that are obtained on the solutions provided they exist. This allows us to find a suitable constant R required for the function g_R in (3.3).

Theorem 3.1 (A priori boundedness of solutions) *Under assumptions (H1)–(H3), there exist constants R, $R_1 > 0$ such that if $u \in W_0^{1,p}(\Omega)$ is any weak solution to (3.1), then*

$$\|u\|_{W_0^{1,p}(\Omega)} \le R_1 \text{ and } \|u\|_{C(\overline{\Omega})} \le R.$$

Proof When we insert $v = u$ in formula (3.5) we obtain what follows

$$\langle A(u), u \rangle = \int_\Omega g(u(x))|\nabla u(x)|^p \, dx - \|\nabla u\|_{L^q(\Omega)}^q$$

$$- \int_\Omega f(x, u(x), \nabla u(x)) u(x) \, dx.$$

From assumption **(H2)** and the variational expression of λ_1 we obtain the estimate

$$\|\nabla u\|_{L^q(\Omega)}^q + \int_\Omega f(x, u(x), \nabla u(x)) u(x) \, dx$$

$$\le |\Omega|^{\frac{p-q}{p}} \|\nabla u\|_{L^p(\Omega)}^q + c_0 \|\nabla u\|_{L^p(\Omega)}^p + c_1 \int_\Omega |u(x)|^\alpha \, dx + c_1 |\Omega|$$

$$\le |\Omega|^{\frac{p-q}{p}} \|\nabla u\|_{L^p(\Omega)}^q + c_0 \|\nabla u\|_{L^p(\Omega)}^p + c_1 |\Omega|^{\frac{p-\alpha}{p}} \|u\|_{L^p(\Omega)}^\alpha + c_1 |\Omega|$$

$$\le |\Omega|^{\frac{p-q}{p}} \|\nabla u\|_{L^p(\Omega)}^q + c_0 \|\nabla u\|_{L^p(\Omega)}^p + \frac{c_1 |\Omega|^{\frac{p-\alpha}{p}}}{\lambda_1^{\alpha/p}} \|\nabla u\|_{L^p(\Omega)}^\alpha + c_1 |\Omega|.$$

Moreover,

$$\int_\Omega g(u(x))|\nabla u(x)|^p \, dx \ge a_0 \|\nabla u\|_{L^p(\Omega)}^p.$$

Summarizing we arrive at

$$(a_0 - c_0) \|\nabla u\|_{L^p(\Omega)}^p \le |\Omega|^{\frac{p-q}{p}} \|\nabla u\|_{L^p(\Omega)}^q + \frac{c_1 |\Omega|^{\frac{p-\alpha}{p}}}{\lambda_1^{\alpha/p}} \|\nabla u\|_{L^p(\Omega)}^\alpha + c_1 |\Omega|. \quad (3.7)$$

Since $c_0 < a_0$, $q < p$ and $\alpha < p$ we see that there exists a constant $R_1 > 0$ such that

$$\|u\|_{W_0^{1,p}(\Omega)} \le R_1.$$

Then by the Sobolev inequality (3.2) we see that

$$\|u\|_{C(\overline{\Omega})} \le R$$

for $R = R_1 \cdot C_S$.   □

3.4 On the Truncated Problem

The next proposition focuses on the properties of the operator (for a given $R > 0$)

$$A_R : W_0^{1,p}(\Omega) \to W^{-1,p'}(\Omega)$$

given by

$$\begin{aligned}
\langle A_R(u), v \rangle = & \int_{\Omega} g_R(u(x))|\nabla u(x)|^{p-2}\nabla u(x) \nabla v(x) \; dx \\
& - \int_{\Omega} |\nabla u|^{q-2}\nabla u(x) \nabla v(x) \; dx \qquad\qquad (3.8) \\
& - \int_{\Omega} f(x, u(x), \nabla u(x))v(x) \; dx
\end{aligned}$$

for $u, v \in W_0^{1,p}(\Omega)$, connected to problem (3.4). For the sake of simplifying the notation we set

$$\langle A_R^1(u), v \rangle = \int_{\Omega} g_R(u(x))|\nabla u(x)|^{p-2}\nabla u(x) \nabla v(x) \; dx$$

and

$$\langle A_R^2(u), v \rangle = - \int_{\Omega} |\nabla u(x)|^{q-2}\nabla u(x) \nabla v(x) \; dx - \int_{\Omega} f(x, u(x), \nabla u(x))v(x) \; dx$$

for $u, v \in W_0^{1,p}(\Omega)$, thus $A_R = A_R^1 + A_R^2$ according to (3.8).

Proposition 3.1 (Properties of truncated operator) *Let $R > 0$ be fixed. Assume that conditions (H1)-(H3) are satisfied. Then the following assertions hold:*

(i) A_R is well defined and bounded;
(ii) A_R^1 has the $(S)_+$ property (recall that it means that any sequence $\{u_n\}_{n\ge1} \subset W_0^{1,p}(\Omega)$ with $u_n \rightharpoonup u$ in $W_0^{1,p}(\Omega)$ and

$$\limsup_{n\to\infty} \langle A_R^1(u_n), u_n - u \rangle \le 0$$

fulfills $u_n \to u$ in $W_0^{1,p}(\Omega)$).
(iii) A_R is continuous.

Proof On the basis of Theorem 1.1 we note that the operator A_R^2 is continuous. It is bounded due to Lemma 3.1 in conjunction with the inequality

$$\|\nabla u\|_{L^q}^q \leq |\Omega|^{\frac{p-q}{p}} \|\nabla u\|_{L^p(\Omega)}^q \quad \text{for all } u \in W_0^{1,p}(\Omega).$$

Now we focus on the operator A_R^1.

By the continuity of g_R we have for all $u, v \in W_0^{1,p}(\Omega)$ that

$$\int_\Omega g_R(u(x))|\nabla u(x)|^{p-1}|\nabla v(x)|\,\mathrm{d}x \leq \int_{\{x:|u(x)|\leq R\}} g_R(u(x))|\nabla u(x)|^{p-1}|\nabla v(x)|\,\mathrm{d}x$$

$$+ \int_{\{x:u(x)>R\}} g_R(R))|\nabla u(x)|^{p-1}|\nabla v(x)|\,\mathrm{d}x$$

$$+ \int_{\{x:u(x)<-R\}} g_R(-R)|\nabla u(x)|^{p-1}|\nabla v(x)|\,\mathrm{d}x$$

$$\leq \max_{t\in[-R,R]} g(t) \int_{\{x:|u(x)|\leq R\}} |\nabla u(x)|^{p-1}|\nabla v(x)|\,\mathrm{d}x$$

$$+ \max_{t\in[-R,R]} g(t) \int_{\{x:|u(x)|>R\}} |\nabla u(x)|^{p-1}|\nabla v(x)|\,\mathrm{d}x$$

$$\leq \max_{t\in[-R,R]} g(t)\|u\|_{W_0^{1,p}(\Omega)}^{p-1}\|v\|_{W_0^{1,p}(\Omega)}.$$

Therefore the operator A_R^1 is well defined and bounded.

Due to the assumption that $p > N$ we can restrict ourselves to the case where $p > 2$. Let $\{u_n\}_{n\geq 1} \subset W_0^{1,p}(\Omega)$ with $u_n \rightharpoonup u$ in $W_0^{1,p}(\Omega)$ and

$$\limsup_{n\to\infty} \int_\Omega g_R(u_n(x))|\nabla u_n(x)|^{p-2}(\nabla u_n(x) - \nabla u(x))\,\mathrm{d}x \leq 0. \tag{3.9}$$

Using the inequality

$$\langle -\Delta_p(u) - (-\Delta_p(v)), u - v\rangle$$

$$= \int_\Omega \left(|\nabla u(x)|^{p-2}\nabla u(x) - |\nabla v(x)|^{p-2}\nabla v(x)\right)(\nabla u(x) - \nabla v(x))\,\mathrm{d}x$$

$$\geq \frac{1}{2^p}\int_\Omega |\nabla u(x) - \nabla v(x)|^p\,\mathrm{d}x$$

for all $u, v \in W_0^{1,p}(\Omega)$, we see that

$$\int_\Omega g_R(u_n(x)) |\nabla u_n(x)|^{p-2} (\nabla u_n(x) - \nabla u(x)) \, dx$$

$$\geq a_0 \int_\Omega \left(|\nabla u_n(x)|^{p-2} \nabla u_n(x) - |\nabla u(x)|^{p-2} \nabla u(x)\right) (\nabla u_n(x) - \nabla u(x)) \, dx$$

$$+ \int_\Omega g_R(u_n(x)) |\nabla u(x)|^{p-2} \nabla u(x) \nabla(u_n - u)(x) \, dx$$

$$\geq \frac{a_0}{2^p} \int_\Omega |\nabla u_n(x) - \nabla u(x)|^p \, dx$$

$$+ \int_\Omega g_R(u_n(x)) |\nabla u(x)|^{p-2} \nabla u(x) \nabla(u_n - u)(x) \, dx.$$

The last integral goes to 0 since u_n converges weakly to u and the function g_R is bounded. Hence by (3.9) we find that $u_n \to u$ in $W_0^{1,p}(\Omega)$ and A_R^1 has the $(S)_+$ property.

In order to check the continuity of A_R^1 let $u_n \to u$ in $W_0^{1,p}(\Omega)$. Passing to a subsequence, the sequence $\{u_n\}$ can be chosen so that it converges a.e. on Ω. For any $v \in W_0^{1,p}(\Omega)$ it holds

$$\left|\langle A_R^1(u_n) - A_R^1(u), v\rangle\right|$$

$$\leq \max_{t \in [-R,R]} g(t) \int_\Omega \left|\left(|\nabla u_n(x)|^{p-2} \nabla u_n(x) - |\nabla u(x)|^{p-2} \nabla u(x)\right) \nabla v(x)\right| \, dx$$

$$+ \int_\Omega |g_R(u_n(x)) - g_R(u(x))| \, |\nabla u(x)|^{p-1} \, |\nabla v(x)| \, dx.$$

Classical arguments provide

$$\int_\Omega \left|\left(|\nabla u_n(x)|^{p-2} \nabla u_n(x) - |\nabla u(x)|^{p-2} \nabla u(x)\right) \nabla v(x) \, dx\right| \to 0 \text{ as } u_n \to u.$$

Since $u_n(x) \to u(x)$ for a.e. $x \in \Omega$, the continuity of g_R entails

$$g_R(u_n(x)) - g_R(u(x)) \to 0 \text{ for a.e. } x \in \Omega.$$

Using

$$|g_R(u_n(x)) - g_R(u(x))| \leq 2 \max_{t \in [-R,R]} g(t),$$

we get through the Lebesgue dominated convergence theorem that

$$\int_{\Omega} |g_R(u_n(x)) - g_R(u(x))| \, |\nabla u(x)|^{p-1} \, |\nabla v(x)| \, dx \to 0 \text{ as } u_n \to u.$$

The continuity of A_R^1 is established. $\qquad\qquad\qquad\qquad\qquad\qquad\qquad\qquad\square$

Remark 3.1 We point out that the operator A_R does not satisfy the $(S)_+$ property. If it did we would arrive at the weak continuity of the $W_0^{1,p}(\Omega)$ norm which is not true.

Concerning the relation between problems (3.1) and (3.4), Theorem 3.1 renders that the shown estimates are independent of the solutions. This fact proves the following result.

Theorem 3.2 *Under assumptions **(H1)–(H3)**, the weak solutions to problems (3.1) and (3.4) coincide.*

3.5 On the Solvability of the Truncated Problem

We now give the definition of a generalized solution.

Definition 3.1 (*Generalized solution*) Assume that hypothesis **(H1)–(H3)** are verified. A function $u \in W_0^{1,p}(\Omega)$ is said to be a generalized solution to problem (3.4) if there exists a sequence $\{u_n\}_{n\geq 1}$ in $W_0^{1,p}(\Omega)$ such that

(a) $u_n \rightharpoonup u$ in $W_0^{1,p}(\Omega)$ as $n \to \infty$;

(b) $\lim_{n\to\infty} \langle A_R(u_n), v \rangle = 0$ for each $v \in W_0^{1,p}(\Omega)$;

(c) $\lim_{n\to\infty} \langle A_R(u_n), u_n - u \rangle = 0$.

Remark 3.2 Writing conditions (b) and (c) explicitly read as

$$-\operatorname{div}\left(g_R(u_n(\cdot))|\nabla u_n(\cdot)|^{p-2}\nabla u_n(\cdot)\right) + \Delta_q u_n(\cdot) - f(\cdot, u_n(\cdot), \nabla u_n(\cdot)) \rightharpoonup 0 \quad \text{in}$$

$W^{-1,p'}(\Omega)$ as $n \to \infty$ and

$$\int_{\Omega} \left(g_R(u_n)|\nabla u_n|^{p-2}\nabla u_n - |\nabla u_n|^{q-2}\nabla u_n\right) \nabla(u_n - u) \, dx$$

$$-\int_{\Omega} f(x, u_n, \nabla u_n)(u_n - u) \, dx \to 0 \text{ as } n \to \infty.$$

By Proposition 3.1 we see that the above definition is well posed. We provide an additional definition.

Definition 3.2 (**Strong generalized solution**) A function $u \in W_0^{1,p}(\Omega)$ is said to be a strong generalized solution to problem (3.4) if there exists a sequence $\{u_n\}_{n\geq 1}$ in $W_0^{1,p}(\Omega)$ such that (a) and (b) in Definition 3.1 are satisfied together with the following condition called (c)'

$$\lim_{n \to \infty} \left\langle A_R^1 \left(u_n\right) + \Delta_q u_n, u_n - u \right\rangle = 0.$$

Since the Banach space $W_0^{1,p}(\Omega)$ with $1 < p < +\infty$ is separable, we can fix a Galerkin base of $W_0^{1,p}(\Omega)$, which actually means a sequence $\{E_n\}_{n \geq 1}$ of finite dimensional vector subspaces of $W_0^{1,p}(\Omega)$ satisfying

(i) $\dim(E_n) < \infty, \quad \forall n$;
(ii) $E_n \subset E_{n+1}, \quad \forall n$;
(iii) $\overline{\bigcup_{n=1}^{\infty} E_n} = W_0^{1,p}(\Omega)$.

An abstract Galerkin scheme has already been introduced in Remark 2.11.

Proposition 3.2 *Assume that conditions (**H2**)-(**H3**) are fulfilled. Then for each $n \geq 1$ there exists $u_n \in E_n$ such that*

$$\langle A_R \left(u_n\right), v \rangle = 0 \quad \text{for all } v \in E_n. \tag{3.10}$$

Moreover, the sequence $\{u_n\}_{n \geq 1}$ is bounded in $W_0^{1,p}(\Omega)$.

Proof For each $n \geq 1$ by $A_{n,R}$ we denote the restriction of A_R to the subspace E_n, so

$$A_{n,R} : E_n \to E^*.$$

We now obtain

$$(a_0 - c_0) \|\nabla v\|_{L^p(\Omega)}^p \leq |\Omega|^{\frac{p-q}{p}} \|\nabla v\|_{L^p(\Omega)}^q + \frac{c_1 |\Omega|^{\frac{p-\alpha}{p}}}{\lambda_1^{\alpha/p}} \|\nabla v\|_{L^p(\Omega)}^\alpha + c_1 |\Omega| \tag{3.11}$$

for all $v \in E_n$. Using that $p > q$, $p > \alpha$ and $c_0 < a_0$ we infer that if $R > 0$, which is independent of n, is sufficiently large then

$$\langle A_{n,R}(v), v \rangle \geq 0 \text{ whenever } v \in E_n \text{ with } \|\nabla v\|_{L^p(\Omega)} = R.$$

By virtue of Lemma 2.12 with $X = E_n$ and $A = A_{R,n}$ we see that there is $u_n \in E_n$ solving (3.10). Since such u_n satisfies estimate (3.11) we get the second assertion as well. $\qquad\square$

We now proceed to the existence result.

Theorem 3.3 (Existence of generalized solutions) *Assume that conditions (**H1**)-(**H3**) hold. Then there exists a generalized solution to problem (3.4).*

Proof We carry out the proof with the sequence $\{u_n\}_{n \geq 1}$ obtained in Proposition 3.2. This sequence is bounded in $W_0^{1,p}(\Omega)$ and thus contains a weakly convergent subsequence (which we do not renumber) with a weak limit $u \in W_0^{1,p}(\Omega)$. Thus we have condition (a) of Definition 3.1 satisfied. Using Proposition 3.1 we arrive at the conclusion that the sequence required in condition (b) is weakly convergent (possibly up to a subsequence which we again do not renumber). We denote by $\eta \in W^{-1,p'}(\Omega)$ the weak limit obtained.

We show that $\eta = 0$. Let $v \in \bigcup_{n \geq 1} E_n$. There is an integer $m \geq 1$ such that $v \in E_m$. Applying Proposition 3.1, we see that equality (3.10) holds true for all $n \geq m$. Letting $n \to \infty$ in (3.10) it follows that

$$\langle \eta, v \rangle = 0 \text{ for all } v \in \bigcup_{n \geq 1} E_n.$$

Since $\bigcup_{n \geq 1} E_n$ is dense in $W_0^{1,p}(\Omega)$ we conclude that $\eta = 0$. This provides that condition (b) is satisfied.

Now we turn to condition (c). Since $\eta = 0$ we see at once that

$$\lim_{n \to \infty} \langle A_R(u_n), u \rangle = 0. \tag{3.12}$$

Inserting $v = u_n$ in (3.10) provides that $\langle A_R(u_n), u_n \rangle = 0$ for all $n \in \mathbb{N}$, whence

$$\lim_{n \to \infty} \langle A_R(u_n), u_n \rangle = 0. \tag{3.13}$$

Combining (3.12) and (3.13) provides that condition (c) is also satisfied. Therefore u is a generalized solution to (3.4). $\qquad\qquad\square$

The existence of a strong generalized solution will be obtained under a hypothesis strengthening **(H2)**.

(H4) *There exist constants* $c_1 \geq 0, c_2 \geq 0, r \in [1, p^\star), r_1 \in [1, p^\star), r_2 \in [1, p)$ *and a nonnegative function* $\sigma \in L^{r'}(\Omega)$ *such that*

$$|f(x, s, \xi)| \leq \sigma(x) + c_1 |s|^{\frac{p^*}{r_1}} + c_2 |\xi|^{\frac{p}{r_2}} \quad \text{for a.e. } x \in \Omega, \text{ all } s \in \mathbb{R}, \xi \in \mathbb{R}^N.$$

The following lemma can be found in [51].

Lemma 3.2 *Assume that condition **(H4)** is satisfied. Then for any sequence* $\{u_n\}_{n \geq 1} \subset W_0^{1,p}(\Omega)$ *such that* $u_n \rightharpoonup u$ *in* $W_0^{1,p}(\Omega)$ *it holds*

$$\lim_{n \to \infty} \int_\Omega f(x, u_n, \nabla u_n)(u_n - u)\, dx = 0.$$

Theorem 3.4 (Existence of strong generalized solutions) *Assume that conditions **(H1)**, **(H3)**, **(H4)** hold. Then there exists a strong generalized solution to problem (3.4).*

Proof Condition **(H4)** implies condition **(H2)**. Thus by Theorem 3.3 we deduce the existence of a generalized solution u. We need to show that

$$\lim_{n \to \infty} \left\langle A_R^1 (u_n) + \Delta_q u_n, u_n - u \right\rangle = 0. \tag{3.14}$$

From Definition 3.1, condition (c), we know that

$$\lim_{n \to \infty} \left\langle A_R (u_n), u_n - u \right\rangle = 0,$$

which by Lemma 3.2 implies (3.14). $\qquad\qquad\square$

3.6 Abstract Existence Result

We begin with an abstract setting and proceed to an abstract result concerning the existence of a generalized solution, although not a strong generalized solution. In our reasoning, we follow the theoretical approach used in the proof of the Browder–Minty Theorem, taking into account that the monotonicity assumption is omitted. For the abstract result, we assume that E is a separable reflexive Banach space and let $A : E \to E^*$. For any fixed $f \in E^*$ we consider the problem

$$A (u) = f. \tag{3.15}$$

Definition 3.3 (*Abstract generalized solution*) An element $u \in E$ is said to be a generalized solution to problem (3.15) if there exists a sequence $\{u_n\}_{n \geq 1}$ in E such that

(a) $u_n \rightharpoonup u$ in E as $n \to \infty$;
(b) $\lim_{n \to \infty} \langle A (u_n) - f, v \rangle = 0$ for each $v \in E$;
(c) $\lim_{n \to \infty} \langle A (u_n), u_n - u \rangle = 0$.

We recall that an operator $A : E \to E^*$ satisfies condition (S) if the relations

$$u_n \rightharpoonup u_0 \text{ in } E$$

and

$$\langle A(u_n), u_n - u_0 \rangle \to 0$$

imply that

$$u_n \to u_0 \text{ in } E.$$

As mentioned in Remark 2.11, since E is separable it contains a dense and countable set $\{h_1, ..., h_n, ...\}$. When we define E_n for $n \in \mathbb{N}$ as the linear hull of $\{h_1, ..., h_n\}$, then the sequence of subspaces E_n has the approximation property: for each $u \in E$

there is a sequence $\{u_n\}_{n\geq 1}$ such that $u_n \in E_n$ for $n \in \mathbb{N}$ and $u_n \to u$. Moreover, $E_n \subset E_{n+1}$ for $n \in \mathbb{N}$ and $\overline{\bigcup_{n=1}^{\infty} E_n} = E$.

Theorem 3.5 (Abstract existence result) *Assume that $A : E \to E^*$ is a continuous, coercive and bounded operator. Then problem (3.15) has at least one generalized solution. If additionally A satisfies the condition (S), then any generalized solution is a weak solution.*

Proof Let us fix $n \in \mathbb{N}$ and the subspace E_n from Remark 2.11. By f_n we denote the restriction of the functional f to the subspace E_n and by A_n the restriction of A to E_n. Then

$$A_n : E_n \to E_n^*$$

is continuous and coercive. Due to the coercivity of A we have that there is some sufficiently large $R > 0$, independent of n, such that

$$\langle A_n (v) - f_n, v \rangle \geq 0 \text{ whenever } v \in E_n \text{ with } \|v\|_E = R.$$

By Lemma 2.12 we now see that the equation

$$A_n (u) = f_n \tag{3.16}$$

has at least one solution u_n with $\|u_n\|_E \leq R$. Since the sequence $\{u_n\}_{n\geq 1}$ is bounded, there is a subsequence, which we do not renumber, convergent weakly to some u. Thus we have shown that condition (a) from Definition 3.3 is satisfied.

Using that the operator A is bounded it follows that $\|A_n (u_n)\|_* \leq M_1$ for some constant $M_1 > 0$ and for all $n \in \mathbb{N}$. We infer from (3.16), again possibly for a subsequence, that

$$\lim_{n \to +\infty} \langle A_n (u_n) , h \rangle = \langle f, h \rangle \text{ for each } h \in \bigcup_{n=1}^{\infty} E_n.$$

By density and since $A_n (u_n) = A (u_n)$ we get that condition (b) from Definition 3.3 is satisfied. This also implies that $A (u_n) \rightharpoonup f$. Testing (3.16) against u_n we have

$$\langle A (u_n) , u_n \rangle = \langle f, u_n \rangle .$$

Since $\lim_{n \to +\infty} \langle A (u_n) - f, u \rangle = 0$ we observe that condition (c) from Definition 3.3 also holds.

Now assume that A satisfies condition (S). We see from $u_n \rightharpoonup u$ and condition (c) that $u_n \to u$ in E. From the definition of A_n and (3.16) we note that for all $n \in \mathbb{N}$ it holds

$$A (u_n) = f_n.$$

By the continuity of A we now obtain that $A (u) = f$ and the assertion follows. $\square$

From the above, it is clear that a generalized solution becomes a weak one on assumption of condition (S). These solutions coincide. The converse result is always valid.

Remark 3.3 In Theorem 3.5 instead of the condition (S), we can assume any other compactness condition introduced in Sect. 2.3. The proof will be nearly identical, except for some technical details.

Proposition 3.3 *Assume that $u \in E$ is a weak solution to (3.15). Then it is also a generalized solution.*

Proof If $u \in E$ is a weak solution to (3.15), we set $\{u_n\}_{n\geq 1} = \{u\}_{n\geq 1}$ and observe that conditions (a), (b), (c) from Definition 3.3 are satisfied. $\qquad\square$

Now we proceed to some immediate applications of Theorem 3.5. In line with Proposition 3.1 and Theorem 3.1 we have that Theorem 3.5 implies Theorem 3.3, whereas from Proposition 3.3 we see that

Lemma 3.3 *Assume that conditions **(H1)-(H3)** are satisfied. Assume that $u \in W_0^{1,p}(\Omega)$ is a weak solution to (3.4). Then it is also a generalized solution.*

Remark 3.4 We note that we cannot say that even a strong generalized solution to (3.4) is a weak solution. Indeed, if u is a strong generalized solution we arrive at the following:

(i) $u_n \rightharpoonup u$ in $W_0^{1,p}(\Omega)$ as $n \to \infty$;

(ii) $\lim_{n\to\infty} \left[\int_\Omega \left(g_R(u_n(x))|\nabla u_n|^{p-2}\nabla u_n - |\nabla u_n|^{q-2}\nabla u_n \right) \nabla (u_n - u) \, dx \right] = 0.$

Hence for u to be a weak solution, it would mean that $u_n \to u$ which is impossible.

Remark 3.5 Due to the type of definition of a strong generalized solution, for the time being there is no abstract counterpart of it.

Finally, we consider the problem with relaxed assumption **(H3)**. Namely, we assume about the nonlinearity that

(H3a) There exist constants $c_0 < a_0$, $c_1/(\lambda_1)^p < a_0 - c_0$ such that

$$f(x,s,\xi)s \leq c_0|\xi|^p + c_1 \left(|s|^p + 1 \right) \text{ for a.e.} x \in \Omega, \text{ all } s \in \mathbb{R}, \xi \in \mathbb{R}^N.$$

Example 3.2 The function $f : \Omega \times \mathbb{R} \times \mathbb{R}^N \to \mathbb{R}$ given by

$$f(x,s,\xi) = |s|^{p-2}s + \frac{s}{1+s^2} \left(|\xi|^{p-1} + h(x) \right) \text{ for all } (x,s,\xi) \in \Omega \times \mathbb{R} \times \mathbb{R}^N,$$

with some $h \in L^\infty(\Omega)$ satisfies conditions **(H2)**, **(H3a)**.

We readily see that Theorem 3.1 holds under assumptions **(H1)**, **(H2)**, **(H3a)**. Thus it is straightforward to obtain

Theorem 3.6 *Assume that conditions **(H1)**, **(H2)**, **(H3a)** hold. Then there exists a generalized solution to problem (3.4).*

3.7 Applications of Abstract Result to Problems with Pseudomonotone Operators

We apply Theorem 3.5 to examine the solvability of the following problem

$$\begin{cases} -\operatorname{div}\left(g(u\,(x))|\nabla u\,(x)\,|^{p-2}\nabla u\,(x) + |\nabla u\,(x)\,|^{q-2}\nabla u\,(x)\right) \\ = f(x, u\,(x)\,, \nabla u\,(x)) & \text{in } \Omega, \\ u\,(x) = 0 & \text{on } \partial\Omega. \end{cases} \qquad (3.17)$$

Recall that **(H4)** implies **(H2)**. Note that due to assumption **(H1)**, problem (3.17) is of independent interest since it cannot be treated directly. For this problem we obtain that the generalized, the strong generalized and the weak solutions coincide. With basically the same proof as for Theorem 3.1 we obtain:

Theorem 3.7 *Assume that conditions **(H1)**, **(H3)**, **(H4)** are satisfied. Then there exists a constant $R > 0$ such that for each weak solution $u \in W_0^{1,p}(\Omega)$ to problem (3.17) it holds the uniform estimate $\|u\|_{C(\overline{\Omega})} \le R$. The constant R depends on g only through its lower bound a_0.*

Further we consider a truncated counterpart of (3.17) with g_R defined by (3.3):

$$\begin{cases} -\operatorname{div}\left(g_R(u\,(x))|\nabla u\,(x)\,|^{p-2}\nabla u\,(x) + |\nabla u\,(x)\,|^{q-2}\nabla u\,(x)\right) \\ = f(x, u\,(x)\,, \nabla u\,(x)) & \text{in } \Omega, \\ u\,(x) = 0 & \text{on } \partial\Omega \end{cases} \qquad (3.18)$$

We define A_R^1, $A_R^2 : W_0^{1,p}(\Omega) \to W^{-1,p'}(\Omega)$ by

$$\langle A_R^1(u), v\rangle = \int_\Omega g_R(u\,(x))|\nabla u\,(x)\,|^{p-2}\nabla u\,(x)\,\nabla v\,(x)\,\mathrm{d}x$$
$$+ \int_\Omega |\nabla u(x)|^{q-2}\nabla u\,(x)\,\nabla v\,(x)\,\mathrm{d}x$$

and

$$\langle A_R^2(u), v\rangle = \int_\Omega f(x, u\,(x)\,, \nabla u\,(x))w\,(x)\,\mathrm{d}x$$

for all $u, v \in W_0^{1,p}(\Omega)$. Then we set $\widetilde{A}_R = A_R^1 + A_R^2$. The operator $\widetilde{A}_R$ has the following properties:

Proposition 3.4 *Let $R > 0$ be fixed. Assume that conditions (H1), (H3), (H4) are satisfied. Then the following assertions hold:*

(i) *$\widetilde{A}_R$ is well defined and bounded;*

(ii) *$\widetilde{A}_R$ has the (S_+) property.*

(iii) *$\widetilde{A}_R$ is continuous.*

Proof Properties (i) and (iii) are immediate from Proposition 3.1. Let $\{u_n\}_{n\geq 1} \subset W_0^{1,p}(\Omega)$ with $u_n \rightharpoonup u$ in $W_0^{1,p}(\Omega)$ and

$$\limsup_{n\to\infty}\left\langle \widetilde{A}_R\left(u_n\right), u_n - u\right\rangle \leq 0$$

Lemma 3.2 renders

$$\limsup_{n\to\infty}\left\langle A_R^1\left(u_n\right), u_n - u\right\rangle \leq 0.$$

Notice the estimates

$$\left\langle A_R^1\left(u_n\right), u_n - u\right\rangle = \int_\Omega g_R\left(u_n(x)\right)|\nabla u_n(x)|^{p-2}\nabla u_n(x)\nabla(u_n - u)(x)\,\mathrm{d}x$$

$$+ \int_\Omega |\nabla u_n(x)|^{q-2}\nabla u_n(x)\nabla(u_n - u)(x)\,\mathrm{d}x$$

$$\geq a_0 \int_\Omega \left(|\nabla u_n|^{p-2}\nabla u_n - |\nabla u|^{p-2}\nabla u\right)\nabla(u_n - u)\,\mathrm{d}x$$

$$+ \int_\Omega g_R\left(u_n\right)|\nabla u|^{p-2}\nabla u\nabla(u_n - u)\,\mathrm{d}x$$

$$+ \int_\Omega \left(|\nabla u_n|^{q-2}\nabla u_n - |\nabla u|^{q-2}\nabla u\right)\nabla(u_n - u)\,\mathrm{d}x$$

$$+ \int_\Omega |\nabla u(x)|^{q-2}\nabla u(x)\nabla(u_n - u)(x)\,\mathrm{d}x$$

$$\geq \frac{a_0}{2^p}\int_\Omega |\nabla u_n(x) - \nabla u(x)|^p\,\mathrm{d}x$$

$$+ \int_\Omega g_R\left(u_n(x)\right)|\nabla u(x)|^{p-2}\nabla u(x)\nabla(u_n - u)(x)\,\mathrm{d}x$$

$$+ \int_\Omega |\nabla u(x)|^{q-2}\nabla u(x)\nabla(u_n - u)(x)\,\mathrm{d}x,$$

from which we have that $u_n \to u$ in $W_0^{1,p}(\Omega)$. This proves properties (ii), thus completing the proof. $\qquad\square$

Now we proceed to resolving (3.17).

Theorem 3.8 *Assume that conditions **(H1)**, **(H3)**, **(H4)** are satisfied. Then problem (3.17) has at least one bounded solution.*

Proof Since condition (S_+) implies condition (S), by Proposition 3.4 we see that Theorem 3.5 can be applied to problem (3.18) which thus has at least one weak solution. Taking into account that any solution to (3.18) with $R > 0$ sufficiently large solves (3.17) too, we arrive at the stated conclusion. $\qquad\square$

3.8 The Case of $1 < p < N$ in the Model Problem

The assumption that $p > N$ was required due to the presence of the weight $g(u(x))$. Now we consider the problem with competing (p, q)-Laplacian but without the weight $g(u(x))$, that is,

$$\begin{cases} - \operatorname{div}\left(|\nabla u\,(x)\,|^{p-2}\nabla u\,(x)\right) + \operatorname{div}\left(|\nabla u\,(x)\,|^{q-2}\nabla u\,(x)\right) \\ = f(x, u\,(x)\,, \nabla u\,(x)) & \text{in } \Omega, \\ u\,(x) = 0 & \text{on } \partial\Omega. \end{cases} \qquad (3.19)$$

For problem (3.19) we will undertake the existence of generalized solutions under the following assumptions which differ from those employed above due to the different space setting:

(H5)   There exist a nonnegative function $\sigma \in L^{(p^*)'}(\Omega)$ and constants $b \geq 0$ and $c \geq 0$ such that

$$|f(x, s, \xi)| \leq \sigma(x) + b|s|^{p^*-1} + c|\xi|^{p-1} \quad \text{for a.e. } x \in \Omega, \text{ all } s \in \mathbb{R}, \xi \in \mathbb{R}^N.$$

Here $(p^*)' = \frac{p^*}{p^*-1}$.

(H6)   There exist constants $c_0 < 1, c_1 > 0$ and $\alpha \in [1, p)$ such that

$$f(x, s, \xi)s \leq c_0|\xi|^p + c_1 (|s|^\alpha + 1) \text{ for a.e. } x \in \Omega, \text{ all } s \in \mathbb{R}, \xi \in \mathbb{R}^N.$$

We provide a simple example of function verifying **(H5)**-**(H6)**:

Example 3.3 The function $f : \Omega \times \mathbb{R} \times \mathbb{R}^N \to \mathbb{R}$ given by

$$f(x, s, \xi) = |s|^{\alpha-2}s + \frac{s}{1 + s^2} \left(|\xi|^{p-1} + h(x)\right) \text{ for all } (x, s, \xi) \in \Omega \times \mathbb{R} \times \mathbb{R}^N,$$

with a constant $\alpha \in [1, p)$ and some $h \in L^\infty(\Omega)$, satisfies conditions **(H5)**-**(H6)**.

We list the following results.

Lemma 3.4 *Under assumption **(H5)** there is a constant $C > 0$ such that*

$$\left| \int_{\Omega} f(x, u(x), \nabla u(x)) v(x)\, dx \right|$$

$$\leq \int_{\Omega} |f(x, u(x), \nabla u(x)) v(x)|\, dx \tag{3.20}$$

$$\leq C \left(\|\sigma\|_{L^{(p^*)'}(\Omega)} + \|u\|_{L^{p^*}(\Omega)}^{p^*-1} + \|\nabla u\|_{L^p(\Omega)}^{p-1} \right) \|\nabla v\|_{L^p(\Omega)}$$

for all $u, v \in W_0^{1,p}(\Omega)$. Moreover, the Niemytskij operator $N_f : W_0^{1,p}(\Omega) \to W^{-1,p'}(\Omega)$ induced by f, namely,

$$N_f(w) = f(\cdot, w(\cdot), \nabla w(\cdot)) \quad \text{for all } w \in W_0^{1,p}(\Omega)$$

is well defined, bounded and continuous.

Proof Using the Hölder inequality and assumption **(H5)** we readily obtain estimate (3.20). Therefore, by the Sobolev embedding theorem (see Section 1.2), there exists a constant $C > 0$ such that

$$\|N_f(w)\|_{W^{-1,p'}(\Omega)} \leq C \left(\|\sigma\|_{L^{(p^*)'}(\Omega)} + \|w\|_{L^{p^*}(\Omega)}^{p^*-1} + \|\nabla w\|_{L^p(\Omega)}^{p-1} \right)$$

for all $w \in W_0^{1,p}(\Omega)$. $\square$

We introduce the mapping $A : W_0^{1,p}(\Omega) \to W^{-1,p'}(\Omega)$ by

$$\langle A(u), v \rangle = \langle -\Delta_p u + \Delta_q u, v \rangle - \int_{\Omega} f(x, u(x), \nabla u(x)) v(x)\, dx \tag{3.21}$$

for all $u, v \in W_0^{1,p}(\Omega)$.

Lemma 3.5 *Under assumptions **(H5)**–**(H6)** the operator A defined by (3.21) is coercive and continuous.*

Proof Through hypothesis (H6), the Hölder inequality and Sobolev embedding theorem we are able to write

$$\langle A(u), u \rangle = \|\nabla u\|_{L^p(\Omega)}^p - \|\nabla u\|_{L^q(\Omega)}^q + \int_{\Omega} f(x, v, \nabla u)\, u\, dx$$

$$\geq \|\nabla u\|_{L^p(\Omega)}^p - |\Omega|^{p-q} \|\nabla u\|_{L^p(\Omega)}^q - c_0 \|\nabla u\|_{L^p(\Omega)}^p - \tilde{c}_1 \left(\|\nabla u\|_{L^p(\Omega)}^{\alpha} + 1 \right),$$

with a positive constant $\tilde{c}_1$. From this estimate, recalling that $p > q$, $p > \alpha$ and $c_0 < 1$ we achieve the desired conclusion about the coercivity. Lemma 3.4 allows us to reach the conclusion about the continuity of the operator A. $\square$

In view of the above results and taking into account Theorem 3.5 we arrive at the following conclusion.

Theorem 3.9 *Under assumptions (H5)-(H6) problem (3.19) has at least one generalized solution.*

In order to get a strong generalized solution we need to strengthen the growth condition in **(H5)** to that of **(H4)**.

Remark 3.6 Condition **(H4)** implies condition **(H5)** because if $r_1 \in [1, p^*)$ and $r_2 \in [1, p\,)$, then $r_1' > (p^*)'$ and $r_2' > p'$, which yields

$$\frac{p^\star}{r_1'} < \frac{p^\star}{(p^\star)'} = p^\star - 1 \text{ and } \frac{p}{r_2'} < \frac{p}{p'} = p - 1.$$

Theorem 3.10 *Under assumptions (H4) and (H6) problem (3.19) has at least one strong generalized solution.*

Proof As already noted, condition **(H4)** is stronger than condition **(H5)**. Consequently, it is permitted to apply Theorem 3.9 which enables us to find a generalized solution $u \in W_0^{1,p}(\Omega)$. If $\{u_n\}_{n \geq 1}$ is the corresponding sequence of approximate solutions, for having that u is a strong generalized solution it is sufficient to show that

$$\lim_{n \to \infty} \int_\Omega f(x, u_n, \nabla u_n)(u_n - u)\, dx = 0.$$

The latter is guaranteed by Lemma 3.2, which completes the proof. $\square$

More details can be found in [33]. For further developements we refer to [3, 52, 59, 66–69]

3.9 Problems with Mixed Local and Non-local Operators

Fix $s \in (0, 1)$ and $1 < q < p < N$. We formulate the problem

$$\begin{cases} -\Delta_p u - (-\Delta)_q^s u = f(x, u, \nabla u) & \text{in } \Omega, \\ u = 0 & \text{in } \mathbb{R}^N \setminus \Omega. \end{cases} \tag{3.22}$$

We assume that $f : \Omega \times \mathbb{R} \times \mathbb{R}^N \to \mathbb{R}$ is a Carathéodory function satisfying the conditions:

(H7)   There exist $C_1 > 0$ and $\sigma \in L^{p^*/(p^*-1)}(\Omega)$ such that

$$|f(x, t, \xi)| \leq C_1 \left(|t|^{p^*-1} + |\xi|^{\frac{p}{p^*}(p^*-1)} \right) + \sigma(x)$$

for a.e. $x \in \Omega$ and all $(t, \xi) \in \mathbb{R} \times \mathbb{R}^N$.

(H8)   There are positive constants c_0 and c_1 such that $c_0 + c_1 \lambda_{1,p}^{-1} < 1$ and

$$f(x, t, \xi)t \le c_0 |\xi|^p + c_1(|t|^p + 1)$$

for a.e. $x \in \Omega$ and all $(t, \xi) \in \mathbb{R} \times \mathbb{R}^N$.

Problem (3.22) is considered in the space X which consists of those functions from $W_0^{1,p}(\mathbb{R}^N)$ that are set 0 outside of Ω. Therefore by standard identification we denote X as $W_0^{1,p}(\Omega)$ in what follows bearing in mind how we understand this space. In a consequence we retain a number of embeddings and functional setting properties which we freely use in what follows.

Proposition 1.4 ensures the continuous embeddings

$$W_0^{1,p}(\Omega) \hookrightarrow W_0^{1,q}(\Omega) \hookrightarrow W_0^{s,q}(\Omega). \tag{3.23}$$

Hence, due to (3.23), the differential operator

$$u \mapsto -\Delta_p u - (-\Delta)_q^s u$$

is well posed on $W_0^{1,p}(\Omega)$.

We introduce the operator $A : W_0^{1,p}(\Omega) \to W^{-1,p'}(\Omega)$ by

$$\langle A(u), v \rangle := \int_\Omega |\nabla u|^{p-2} \nabla u \cdot \nabla v \, dx$$

$$- \int_{\mathbb{R}^N \times \mathbb{R}^N} \frac{|u(x) - u(y)|^{q-2}(u(x) - u(y))(v(x) - v(y))}{|x - y|^{N+qs_2}} \, dx \, dy$$

for all $u, v \in W_0^{1,p}(\Omega)$. It turns out that A is bounded and continuous. Such an operator is called mixed local-nonlocal operator as it contains both, the (negative) p-Laplacian and also the fractional q-Laplacian.

Lemma 3.6 *Under assumption **(H7)**, there exists a constant $C > 0$ such that*

$$\left| \int_\Omega f(x, u(x), \nabla u(x)) v(x) \, dx \right| \le C \left(\|u\|_{p*}^{p*} + \|\nabla u\|_p^{p(p*-1)/p*} + \|\sigma\|_{p*/(p*-1)} \right) \|v\|_{1,p}$$

for all $u, v \in W_0^{1,p}(\Omega)$.

Proof Using (H_1), Hölder's inequality and the Sobolev inequality we get

$$\left| \int_\Omega f(x, u(x), \nabla u(x)) v(x)\, dx \right| \leq C_1 \left(\int_\Omega |u|^{p^*-1} |v|\, dx + \int_\Omega |\nabla u|^{\frac{p(p^*-1)}{p^*}} |v|\, dx \right) + \int_\Omega |\sigma| |v|\, dx$$

$$\leq C_1 \left(\|u\|_{p^*}^{p^*-1} \|v\|_{p^*} + \|\nabla u\|_p^{\frac{p(p^*-1)}{p^*}} \|v\|_{p^*} + \|\sigma\|_{p^*/(p^*-1)} \|v\|_{p^*} \right)$$

$$\leq C \left(\|u\|_{p^*}^{p^*-1} + \|u\|_{1,p}^{\frac{p(p^*-1)}{p^*}} + \|\sigma\|_{p^*/(p^*-1)} \right) \|v\|_{1,p}$$

with some constant $C > 0$.											$\square$

Lemma 3.6 implies that the Niemytskii operator

$$\mathcal{N}_f : W_0^{1,p}(\Omega) \to W^{-1,p'}(\Omega)$$

defined by

$$\mathcal{N}_f(u) := f(\cdot, u, \nabla u), \quad \forall u \in W_0^{1,p}(\Omega)$$

is bounded. Krasnoselskii's theorem (see Theorem 1.1) guarantees that it is also continuous. Therefore, the operator

$$\mathcal{A}(u) := A(u) - \mathcal{N}_f(u), \quad u \in W_0^{1,p}(\Omega)$$

is well posed, bounded, and continuous.

The presence of competing operators requires a new notion of solution for (3.22) that complies with Definition 3.1.

Definition 3.4 (*Generalized solution*) Assume that assumption **(H7)** holds. A function $u \in W_0^{1,p}(\Omega)$ is called a generalized solution to problem (3.22) provided there exists a sequence $\{u_n\} \subset W_0^{1,p}(\Omega)$ such that

$$u_n \rightharpoonup u \text{ in } E, \quad \mathcal{A}(u_n) \to 0 \text{ in } W^{-1,p'}(\Omega),$$

and

$$\lim_{n \to +\infty} \left(\langle A(u_n), u_n - u \rangle - \int_\Omega f(\cdot, u_n, \nabla u_n)(u_n - u)\, dx \right) = 0.$$

Proposition 3.5 *Under assumptions* **(H7)-(H8)** *the operator* $\mathcal{A}$ *is coercive.*
Proof By a direct calculation we obtain

$$\langle \mathcal{A}(v), v \rangle = \int_\Omega |\nabla v|^p\, dx - \int_{\mathbb{R}^N \times \mathbb{R}^N} \frac{|v(x) - v(y)|^q}{|x - y|^{N+qs}}\, dx\, dy - \int_\Omega f(\cdot, v, \nabla v) v\, dx$$

$$\geq \|v\|_{1,p}^p - \|v\|_{s_2,q}^q - c_0 \int_\Omega |\nabla v|^p\, dx - c_1 \int_\Omega (|v|^p + 1)\, dx$$

$$\geq \left(1 - c_0 - c_1 \lambda_{1,p}^{-1} \right) \|v\|_{1,p}^p - \tilde{c} \left(\|v\|_{1,p}^q + 1 \right), \quad v \in W_0^{1,p}(\Omega),$$

with a constant $\tilde{c} > 0$. Since $c_0 + c_1\lambda_{1,p}^{-1} < 1$ and $q < p$, the assertion is reached. $\square$

Our main result regarding problem (3.22) follows now immediately from Theorem 3.5 since the operator $\mathcal{A}$ is well posed, bounded, coercive and continuous.

Theorem 3.11 *Assume that conditions **(H7)**-**(H8)** hold. Then problem (3.22) admits a generalized solution in the sense of Definition 3.4.*

Related results can be found in [3, 9, 19, 35, 36, 43].

3.10 Problems with Competing Nonlocal Operators

Let $0 < s_2 < s_1 < 1$ and $1 < q < p < \frac{N}{s_1}$. We consider the problem

$$\begin{cases} (-\Delta)_p^{s_1} u - (-\Delta)_q^{s_2} u = f(x,u) & \text{in } \Omega, \\ u = 0 & \text{in } \mathbb{R}^N \setminus \Omega. \end{cases} \tag{3.24}$$

Here $f : \Omega \times \mathbb{R} \to \mathbb{R}$ is a Carathéodory function satisfying the conditions:

(H9) There exist $C_1 > 0$ and $\sigma \in L^{P_{s_1}^*/(P_{s_1}^*-1)}(\Omega)$ such that

$$|f(x,t)| \le C_1|t|^{P_{s_1}^*} + \sigma(x)$$

for a.e. $x \in \Omega$ and all $t \in \mathbb{R}$.

(H10) There is a constant $c_1 > 0$ such that $c_1\lambda_{1,p,s_1}^{-1} < 1$ and

$$f(x,t)t \le c_1(|t|^p + 1)$$

for a.e. $x \in \Omega$ and all $t \in \mathbb{R}$.

It is known from [4, Lemma 2.6] that the continuous embedding

$$W_0^{s_1,p}(\Omega) \hookrightarrow W_0^{s_2,q}(\Omega) \tag{3.25}$$

holds. Therefore, the differential operator

$$u \mapsto (-\Delta)_p^{s_1} u - (-\Delta)_q^{s_2} u$$

is well posed on $W_0^{s_1,p}(\Omega)$.

Let the map $A : W_0^{s_1,p}(\Omega) \to W^{-s_1,p'}(\Omega)$ be defined by

$$\langle A(u), v \rangle := \int_{\mathbb{R}^N \times \mathbb{R}^N} \frac{|u(x) - u(y)|^{p-2}(u(x) - u(y))(v(x) - v(y))}{|x - y|^{N+ps_1}}\, dx\, dy$$

$$- \int_{\mathbb{R}^N \times \mathbb{R}^N} \frac{|u(x) - u(y)|^{q-2}(u(x) - u(y))(v(x) - v(y))}{|x - y|^{N+qs_2}}\, dx\, dy$$

From [28, Lemma 2.1], we derive that the operator A is bounded and continuous. We call this map a competing nonlocal operator.

Lemma 3.7 *Under assumption (H9), there exists a constant $C > 0$ such that*

$$\left| \int_\Omega f(x, u(x)v(x)\, dx \right| \leq C \left(\|u\|_{P_{s_1}^*}^{P_{s_1}^*} + \|\sigma\|_{P_{s_1}^*/(P_{s_1}^* - 1)} \right) \|v\|_{s_1,p}$$

for all $u, v \in W_0^{s_1,p}(\Omega)$.

Proof Using **(H9)**, Hölder's inequality and part (b) in Proposition 1.4, we infer that

$$\left| \int_\Omega f(x, u(x)v(x)\, dx \right| \leq C_1 \int_\Omega |u|^{P_{s_1}^* - 1}|v|\, dx + \int_\Omega |\sigma||v|\, dx$$

$$\leq C_1 \left(\|u\|_{P_{s_1}^*}^{P_{s_1}^* - 1} \|v\|_{P_{s_1}^*} + \|\sigma\|_{P_{s_1}^*/(P_{s_1}^* - 1)} \|v\|_{P_{s_1}^*} \right)$$

$$\leq C \left(\|u\|_{P_{s_1}^*}^{P_{s_1}^* - 1} + \|\sigma\|_{P_{s_1}^*/(P_{s_1}^* - 1)} \right) \|v\|_{s_1,p}$$

with a constant $C > 0$. $\qquad\square$

The Niemytskii operator

$$\mathcal{N}_f : W_0^{s_1,p}(\Omega) \to W^{-s_1,p'}(\Omega), \quad \mathcal{N}_f(u) := f(\cdot, u)$$

is bounded and continuous by Krasnoselskii's Theorem (see Theorem 1.1). Hence the operator

$$\mathcal{A}(u) := A(u) - \mathcal{N}_f(u), \quad u \in W_0^{s_1,p}(\Omega)$$

is well posed, bounded, and continuous.

Definition 3.5 Assume that hypothesis **(H9)** holds. A function $u \in W_0^{s_1,p}(\Omega)$ is called a generalized solution to problem (3.24) provided there exists a sequence $\{u_n\} \subset W_0^{s_1,p}(\Omega)$ such that

$$u_n \rightharpoonup u \text{ in } W_0^{s_1,p}(\Omega), \quad \mathcal{A}(u_n) \rightharpoonup 0 \text{ in } W^{-s_1,p'}(\Omega),$$

and

$$\lim_{n \to +\infty} \left(\langle A(u_n), u_n - u \rangle - \int_\Omega f(\cdot, u_n)(u_n - u)\, dx \right) = 0.$$

Proposition 3.6 *Under assumptions (H9)-(H10), the operator $\mathcal{A}$ is coercive.*

Proof By **(H10)**, Hölder's inequality and (3.25), we get

$$\langle \mathcal{A}(v), v \rangle = \int_{\mathbb{R}^N \times \mathbb{R}^N} \frac{|v(x) - v(y)|^p}{|x - y|^{N+ps_1}}\, dx\, dy - \int_{\mathbb{R}^N \times \mathbb{R}^N} \frac{|v(x) - v(y)|^q}{|x - y|^{N+qs_2}}\, dx\, dy$$

$$- \int_\Omega f(\cdot, v) v\, dx$$

$$\geq \|v\|_{s_1,p}^p - \|v\|_{s_2,q}^q - c_1 \int_\Omega (|v|^p + 1)\, dx$$

$$\geq \left(1 - c_1 \lambda_{1,p,s_1}^{-1}\right) \|v\|_{s_1,p}^p - C\left(\|v\|_{s_1,p}^q + 1\right), \quad v \in W_0^{s_1,p}(\Omega)$$

with a constant $C > 0$. The conditions $c_1 \lambda_{1,p,s_1}^{-1} < 1$ and $q < p$ imply the assertion.
$\square$

Our main result regarding problem (3.24) follows now immediately from Theorem 3.5 since the operator $\mathcal{A}$ is well posed, bounded, coercive, and continuous.

Theorem 3.12 (Existence of generalized solutions) *Assume that hypotheses (H9)-(H10) hold. Then problem (3.24) admits a generalized solution (in the sense of Definition 3.5).*

Chapter 4
Generalized Solutions—Variational Problems

Abstract In this chapter we are interested in equations given in potential form, i.e. equations that arise from equating to zero the Gâteaux derivative of certain Euler action functionals. When the action functional is continuous and coercive but lacks sequential weak lower semicontinuity, we cannot apply the Weierstrass–Tonelli theorem, but we can still obtain the existence of finite-dimensional approximations and investigate their convergence. Thereby, we obtain a new notion of variational generalized solution, which is introduced in this chapter in an abstract form and then illustrated by the solvability of problems involving the competing (p, q)-Laplacian with unbounded weight in the leading term.

Keywords Competing (p, q)-Laplacian; potential operator · Ritz approximations · Variational abstract generalized solution · Variational generalized solution.

4.1 Abstract Existence Result

An abstract statement given in the previous chapter asserts for a continuous, coercive and bounded operator $A : E \to E^*$ on a separable and reflexive Banach space E that the equation

$$A(u) = h \tag{4.1}$$

has at least one generalized solution $u \in E$ for each $h \in E^*$, in the sense that there exists a sequence $\{u_n\}_{n \geq 1}$ in E such that

(a) $u_n \rightharpoonup u$ in E as $n \to \infty$;
(b) $\lim_{n \to \infty} \langle A(u_n) - h, v \rangle = 0$ for each $v \in E$;
(c) $\lim_{n \to \infty} \langle A(u_n), u_n - u \rangle = 0$.

In case where the operator A satisfies condition (S) or related, see Sect. 2.3 any generalized solution becomes a weak solution, that is, $\langle A(u) - h, v \rangle = 0$ for all $v \in E$. Inspired by the preceding idea, we introduce, after [33], a corresponding notion of generalized variational solution regarding (4.1) when A is a potential operator

M. Galewski and D. Motreanu, *Competing Operators and Their Applications to Boundary Value Problems*, SpringerBriefs in Mathematics,
https://doi.org/10.1007/978-3-032-15445-3_4

and investigate its existence. This time the existence result is not based on resolving operator equations but on a minimization approach. Significant differences arise in the reasoning. For instance, we do not do assume that the operator is coercive, but that occurs for its potential. We emphasize that the coercivity of the potential of a mapping need not imply the coercivity of the mapping. For example, here is a coercive function

$$f(x) = \begin{cases} x^2 - 1, & x \le 2 \\ 4x - 5, & x > 2 \end{cases}$$

whose derivative is not coercive. This phenomenon motivates the application of variational methods instead of methods of monotone operators at the expense of imposing the potentiality assumption on the terms appearing in the equation.

We start with a minimization setting related to a Galerkin type scheme introduced in Remark 2.11 and recalled prior to the definition of the abstract generalized solution, see Definition 3.3. For the abstract result we assume that E is a separable reflexive Banach space. We recall that a functional $J : E \to \mathbb{R}$ is coercive on a Banach space E if $\lim_{\|u\| \to +\infty} J(u) = +\infty$. Since E is separable, we fix a Galerkin basis of E meaning a sequence of finite-dimensional subspaces E_n such that $E_n \subset E_{n+1}$ for each $n \in \mathbb{N}$ and $\overline{\bigcup_{n=1}^{\infty} E_n} = E$.

Theorem 4.1 *Assume that the functional $J : E \to \mathbb{R}$ is coercive on E, lower semicontinuous on each E_n, and Gâteaux differentiable on E with differential $J' : E \to E^*$ being a bounded operator. Then there exist some $u_0 \in E$ and a sequence $\{u_n\}_{n \ge 1}$ such that $u_n \in E_n$ for each $n \in \mathbb{N}$,*

$$\inf_{u \in E_n} J(u) = J(u_n) \to \inf_{u \in E} J(u),$$

$u_n \rightharpoonup u_0$ in E, $J'(u_n) \rightharpoonup 0$ in E^*, and $\langle J'(u_n), u_n - u_0 \rangle \to 0$ as $n \to \infty$.

Proof The functional $J_n := J|_{E_n}$ is lower semicontinuous and coercive on E_n. Since E_n is finite dimensional, there exists $u_n \in E_n$ such that

$$\inf_{u \in E_n} J(u) = J(u_n) = J_n(u_n) \text{ and } J'_n(u_n) = 0.$$

Taking into account that J is coercive, we obtain that the sequence $\{u_n\}_{n \ge 1}$ is bounded and thus, since E is reflexive, $u_n \rightharpoonup u_0$ in E with some $u_0 \in E$, up to a subsequence. Using that J' is a bounded operator, there is $\eta \in E^*$ with $J'(u_n) \rightharpoonup \eta$ in E^*, up to a subsequence.

Let $v \in E_m$ for some m. It follows that $\langle J'(u_n), v \rangle = 0$ for all $n \ge m$, so $\langle \eta, v \rangle = 0$. The equality

$$\overline{\bigcup_{n=1}^{\infty} E_n} = E \tag{4.2}$$

entails $\eta = 0$, thus

$$J'(u_n) \rightharpoonup 0 \text{ in } E^* \text{ and } \langle J'(u_n), u_n - u_0 \rangle \to 0.$$

The second convergence in the above follows since $\langle J'(u_n), u_n \rangle = 0$ for all $n \in \mathbb{N}$. The convergence

$$\inf_{u \in E_n} J(u) \to \inf_{u \in E} J(u)$$

is a consequence of (4.2). This completes the proof. $\square$

Here is a result addressing the convergence of a minimizing sequence which is in fact a version of Weierstrass-Tonelli Theorem.

Corollary 4.1 *Assume in addition to the hypotheses of Theorem 4.1 that the differential $J' : E \to E^*$ is continuous and satisfies the condition (S). Then there exist some $u_0 \in E$ and a sequence $\{u_n\}_{n \geq 1}$ such that $u_n \in E_n$ for each $n \in \mathbb{N}$, $u_n \to u_0$ in E,*

$$\inf_{u \in E_n} J(u) = J(u_n) \to J(u_0) = \inf_{u \in E} J(u)$$

and $J'(u_0) = 0$.

Proof Theorem 4.1 and condition (S) for J' ensure the norm convergence $u_n \to u_0$ in E from which through the continuity of J' it is straightforward to reach the stated conclusion. $\square$

Theorem 4.1 suggests a definition of generalized solution to a nonlinear equation in variational form taking into account the variational structure. Assume $A : E \to E^*$ is a potential operator with the potential $\mathcal{A} : E \to \mathbb{R}$, that is, $\mathcal{A}$ is Gâteaux differentiable and $A = \mathcal{A}'$. For any fixed $h \in E^*$ we consider Eq. (4.1) with $A = \mathcal{A}'$. To this equation it is associated the Euler action functional $J : E \to \mathbb{R}$ defined by

$$J(u) = \mathcal{A}(u) - \langle h, u \rangle \text{ for all } u \in E. \tag{4.3}$$

The functional J is Gâteaux differentiable and its critical points coincide with the weak solutions to (4.1) with $A = \mathcal{A}'$, i.e., any $u_0 \in E$ satisfying $\langle A(u_0), v \rangle = \langle h, v \rangle$ for all $v \in E$. Generally, the derivative $J' : E \to E^*$ does not fulfil condition (S) (so, a compactness condition of Palais-Smale type is not guaranteed for the functional J). This makes us to turn to generalized solutions in a variational sense instead of weak solutions.

Definition 4.1 (*Abstract generalized solution*) An element $u \in E$ is said to be a generalized variational solution to problem (4.1) with $A = \mathcal{A}'$ if there exist a sequence of subspaces $\{E_n\}_{n \geq 1}$ satisfying $E_n \subset E_{n+1}$ for each $n \in \mathbb{N}$ and $\overline{\bigcup_{n=1}^{\infty} E_n} = E$ and a sequence of elements $\{u_n\}_{n \geq 1}$ with $u_n \in E_n$ such that

(a) $u_n \rightharpoonup u$ in E as $n \to \infty$;
(b) $\inf_{v \in E_n} J(v) = J(u_n)$;
(c) $J'(u_n) \rightharpoonup 0$ in E^* and $\langle J'(u_n), u_n - u \rangle \to 0$, with J introduced in (4.3).

We infer the following existence result.

Theorem 4.2 (Existence of generalized variational solutions) *Assume that the operator $A : E \to E^*$ is bounded and potential with a coercive potential $\mathcal{A} : E \to \mathbb{R}$. Then problem (4.1) with $A = \mathcal{A}'$ has at least one generalized variational solution in the sense of Definition 4.1. If additionally A is continuous and satisfies the condition (S), then any generalized variational solution of problem (4.1) is a weak solution.*

Proof The result is the consequence of Theorem 4.1 and Corollary 4.1 applied to the functional J introduced in (4.3). $\qquad\qquad\square$

We conclude this section with some related versions of the Weierstrass-Tonelli result, where the convergence of a minimizing sequence for (weak) solutions is a norm one.

Theorem 4.3 (Norm convergence of minimizing sequences) *Assume that $A : E \to E^*$ satisfies condition (S) and it is potential with a sequentially weakly lower semi-continuous and coercive potential $\mathcal{A}$. Let $h \in E^*$ be fixed. Then there is a weak solution $u_0 \in E$ to the equation (4.1) which minimizes the action functional $J : E \to \mathbb{R}$ defined by*

$$J(u) = \mathcal{A}(u) - \langle h, u \rangle$$

and moreover there is a sequence $\{u_n\}_{n \geq 1} \subset E$, $u_n \to u_0$ such that

$$J(u_n) \to \inf_{u \in E} J(u) \ \text{and} \ J'(u_n) \to 0 \ (\text{in } E^*). \qquad\qquad (4.4)$$

If in addition A is strictly monotone, then the solution is unique. Moreover, the functional J satisfies the (PS) condition.

Proof From the assumptions it follows that the functional J is Gâteaux differentiable, coercive and sequentially weakly lower semicontinuous. Hence it is bounded from below. Moreover, it has at least one global minimizer u_0 which is a critical point, i.e. a solution to (4.1). By Theorem 2.6 there is a minimizing sequence $\{u_n\}_{n \geq 1} \subset E$, $u_n \rightharpoonup u_0$ (the weak convergence follows by coercivity) and which is such that (4.4) holds. We see from $J'(u_n) \to 0$ by writing the derivative explicitly that

$$\langle A(u_n) - A(u_0), u_n - u_0 \rangle \to 0. \qquad\qquad (4.5)$$

Since A satisfies condition (S), we obtain that $u_n \to u_0$.

Let us take a Palais-Smale sequence $\{u_n\}_{n \geq 1} \subset E$, i.e., such a sequence that $J(u_n)$ is bounded and $J'(u_n) \to 0$. Due to the assumptions we can suppose that $u_n \rightharpoonup u_0$, possibly up to a subsequence which we choose and do not renumber. Hence (4.5) holds. Since $J' = A - h$ satisfies condition (S) and since $u_n \rightharpoonup u_0$, we see that $u_n \to u_0$, so the result is proven. $\qquad\qquad\square$

We also mention that checking condition (S) is technically similar to checking the strong convergence of a bounded (PS) sequence.

From the proof of the above result we immediately obtain:

Proposition 4.1 *Assume that a Gâteaux differentiable functional $J : E \to \mathbb{R}$ has a derivative $J' : E \to E^*$ which satisfies condition (S). Then any bounded (PS) sequence for the functional J admits a strongly convergent subsequence.*

Now we are in a position to formulate the result about the existence of minimizers for coercive functionals:

Corollary 4.2 *Assume that the functional $J : E \to \mathbb{R}$ is bounded from below, coercive, Gâteaux differentiable and that its derivative $J' : E \to E^*$ satisfies condition (S). Then there is some $u_0 \in E$ such that*

$$J(u_0) = \inf_{u \in E} J(u).$$

Proof From Theorem 4.3 it follows that the functional J satisfies the (PS) condition. Since it is bounded from below and satisfies the (PS) condition, it necessarily has at least one minimizer. $\square$

Remark 4.1 We may replace condition (S) with conditions $(S)_+$ or $(S)_0$ with retaining the same conclusion in the above.

In the result that follows we show that for a potential problem which can be handled by the Banach fixed point theorem the sequence obtained by the method of successive approximations which converges to the unique solution stands also for a minimizing sequence.

Proposition 4.2 *Let E be a Hilbert space with the scalar product $(\cdot, \cdot)$. Let $N : E \to E$ be a contraction with the unique fixed point $u^* \in E$ (guaranteed by the Banach contraction theorem). If there exists a C^1 functional J such that*

$$J'(u) = u - N(u) \text{ for all } u \in E,$$

then u^* minimizes the functional J, i.e.,

$$J(u^*) = \inf_{u \in E} J(u).$$

Moreover, the sequence $\{u_n\}_{n \geq 1} \subset E$ defined by $u_{n+1} = N(u_n)$ for any $u_0 \in E$ is such that $u_n \to u^*$ and

$$J(u_n) \to \inf_{u \in E} J(u) \text{ and } J'(u_n) \to 0. \tag{4.6}$$

Proof Since N is a contraction we see by a direct calculation that J' is strongly monotone. Furthermore, J is coercive and sequentially weakly lower semicontinuous. It also holds that J is strictly convex. Moreover, since J' is strongly monotone, it satisfies condition (S). Then by Theorem 4.3 there is a minimizer, which is unique by the strict convexity and therefore equal to u^*. Assertion (4.6) follows by the continuity of N. $\square$

4.2 Applications to Boundary Value Problems

The objective here is the study of the following problem with homogeneous Dirichlet boundary condition posed in a variational form:

$$\begin{cases} -\operatorname{div}\left(g(p^{-1}|\nabla u(x)|^p)|\nabla u\,(x)\,|^{p-2}\nabla u\,(x)\right) \\ +\operatorname{div}\left(|\nabla u\,(x)\,|^{q-2}\nabla u\,(x)\right) = f(x,u(x)) & \text{in } \Omega \\ u\,(x) = 0 & \text{on } \partial\Omega, \end{cases} \tag{4.7}$$

where $1 < q < p < +\infty$ and Ω is a bounded domain in $\mathbb{R}^N$ with the Lebesgue measure $|\Omega|$. We assume in what follows the condition:

(A0) $g : \mathbb{R} \to \mathbb{R}$ *is a continuous function for which there are constants $a_g > 0$ and $b_g > 0$ and $a_g \leq g(t) \leq b_g$ for all $t \geq 0$;*

and $f : \Omega \times \mathbb{R} \to \mathbb{R}$ is a Carathéodory function satisfying

(A1) *There exist a nonnegative function $\sigma \in L^{(p^*)'}(\Omega)$ and a constant $b \geq 0$ such that*

$$|f(x,s)| \leq \sigma(x) + b|s|^{p^*-1} \quad \text{for a.e. } x \in \Omega, \text{ all } s \in \mathbb{R}.$$

(A2) *There exists a positive constant c_1 with $c_1 < \frac{a_g \lambda_1}{p}$ such that*

$$F(x,s) := \int_0^s f(x,\tau)d\tau \leq c_1 \left(|s|^p + 1\right) \text{ for a.e. } x \in \Omega, \text{ all } s \in \mathbb{R}.$$

Remark 4.2 In condition **(A2)** we have the additional assumption on the constant c_1. We can get rid of it by assuming what follows:

(A2a) *There exists a positive constant c_1 and $\alpha \in (1, p)$ such that*

$$F(x,s) \leq c_1 \left(|s|^\alpha + 1\right) \text{ for a.e. } x \in \Omega, \text{ all } s \in \mathbb{R}.$$

For example, the function $f : \Omega \times \mathbb{R} \to \mathbb{R}$ given by

$$f(x,s) = h_1|s|^{\alpha-2}s + h_2(x)$$

for all $(x, s) \in \Omega \times \mathbb{R}$, with constants $\alpha \in [1, p)$ and $h_1 > 0$ and a function $h_2 \in L^\infty(\Omega)$, satisfies conditions **(A1)-(A2a)**.

Thanks to the assumption $1 < q < p < +\infty$ there is a continuous embedding $W_0^{1,p}(\Omega) \hookrightarrow W_0^{1,q}(\Omega)$. Therefore the operator in the left-hand side of equation (4.7) is well defined on $W_0^{1,p}(\Omega)$. Although the ellipticity condition is not satisfied, problem (4.7) has a variational structure. Indeed, the driving operator admits the potential

$$u \mapsto \int_{\Omega} (G(p^{-1}|\nabla u(x)|^p) - q^{-1}|\nabla u(x)|^q) \mathrm{d}x, \quad \text{where } G(t) = \int_0^t g(s)\mathrm{d}s.$$

The main difficulty in developing a variational approach lies in the nonconvex structure of the potential. As a consequence, it does not allow the Euler functional associated to problem (4.7) be sequentially weakly lower semicontinuous and which also prevents from using the Browder-Minty Theorem. That is why we conduct a variational approach for resolving problem (4.7) in the sense of generalized variational solutions introduced in Definition 4.1.

Lemma 4.1 *Under assumption (A1), the Niemytskij operator*

$$N_f : W_0^{1,p}(\Omega) \to W^{-1,p'}(\Omega)$$

induced by the Carathéodory function

$$f : \Omega \times \mathbb{R} \to \mathbb{R},$$

namely,

$$N_f(w) = f(\cdot, w(\cdot)), \quad \forall w \in W_0^{1,p}(\Omega),$$

is well defined, continuous and bounded. Moreover, there is a constant $C > 0$ such that

$$\left\| N_f(w) \right\|_{W^{-1,p'}(\Omega)} \le C \left(\|\sigma\|_{L^{(p^*)'}(\Omega)} + \|w\|_{L^{p^*}(\Omega)}^{p^*-1} \right) \quad \text{for all } w \in W_0^{1,p}(\Omega). \tag{4.8}$$

Proof Assumption **(A1)**, Hölder's inequality and Sobolev embedding theorem yield a constant $C > 0$ such that

$$\int_{\Omega} |f(x, w(x))v(x)| \, \mathrm{d}x \le C \left(\|\sigma\|_{L^{(p^*)'}(\Omega)} + \|w\|_{L^{p^*}(\Omega)}^{p^*-1} \right) \|\nabla v\|_{L^p(\Omega)}$$

for all $v, w \in W_0^{1,p}(\Omega)$. It turns out that the map N_f is well defined and satisfies estimate (4.8). The continuity of N_f can be readily shown on the basis of the growth condition in assumption (A1) and Lebesgue dominated convergence theorem. $\qquad \square$

Corresponding to problem (4.7) we introduce the Euler action functional $J : W_0^{1,p}(\Omega) \to \mathbb{R}$ by

$$J(u) = \int_{\Omega} G\left(\frac{1}{p}|\nabla u(x)|^p\right) \mathrm{d}x - \frac{1}{q} \int_{\Omega} |\nabla u(x)|^q \, \mathrm{d}x$$

$$- \int_{\Omega} F(x, u(x)) \, \mathrm{d}x, \quad \forall u \in W_0^{1,p}(\Omega). \tag{4.9}$$

Proposition 4.3 *Assume that conditions (A0), (A1) are fulfilled. Then the functional J in (4.9) is continuously differentiable and its differential $J'(u)$ at any $u \in W_0^{1,p}(\Omega)$ has the expression*

$$\langle J'(u), v \rangle = \int_\Omega g\left(\tfrac{1}{p}|\nabla u(x)|^p\right) |\nabla u(x)|^{p-2} \nabla u(x) \, \nabla v(x) \, dx$$
$$- \int_\Omega |\nabla u|^{q-2} \nabla u(x) \, \nabla v(x) \, dx - \int_\Omega f(x, u(x)) v(x) \, dx, \quad \forall v \in W_0^{1,p}(\Omega).$$

$$(4.10)$$

Proof The boundedness of g implies the existence of a constant $c > 0$ with

$$|G(t)| \leq c(|t| + 1) \text{ for all } t \in \mathbb{R}.$$

Then the functional J is Gâteaux differentiable with its Gâteaux differential J' expressed in (4.10). From Lemma 2.15 it follows that the differential $J' : W_0^{1,p}(\Omega) \to W^{-1,p'}(\Omega)$ is continuous, which completes the proof. $\qquad\square$

Now we regard problem (4.7) as an operator equation in the form of (4.1) with $E = W_0^{1,p}(\Omega)$, $h = 0$, and

$$A(u) = -\operatorname{div}\left(g(p^{-1}|\nabla u|^p)|\nabla u|^{p-2}\nabla u\right) + \operatorname{div}\left(|\nabla u|^{q-2}\nabla u\right) - f(\cdot, u) \quad (4.11)$$

for all $u \in W_0^{1,p}(\Omega)$. We note from (4.10) and (4.11) that $Au = J'(u)$ for all $u \in W_0^{1,p}(\Omega)$, with the functional J defined in (4.9).

Theorem 4.4 (Existence of generalized solutions) *Assume that conditions (A0)–(A2) hold. Then there exists a generalized variational solution to problem (4.7) in the sense of Definition 4.1 for the action functional $J : W_0^{1,p}(\Omega) \to \mathbb{R}$ given in (4.9).*

Proof Proposition 4.3 ensures that the functional J introduced in (4.9) is continuously differentiable with the differential J' given by (4.10). From (4.10) we infer that $J' : W_0^{1,p}(\Omega) \to W^{-1,p'}(\Omega)$ is a bounded operator.

By Hölder's inequality we see that

$$\|\nabla u\|_{L^q(\Omega)}^q \leq |\Omega|^{\frac{p-q}{p}} \|\nabla u\|_{L^p(\Omega)}^q, \quad \forall u \in W_0^{1,p}(\Omega).$$

Through assumption (A2) we get

$$\int_\Omega F(x, u(x)) \leq c_1 \lambda_1^{-1} \|\nabla u\|_{L^p(\Omega)}^p + c_1 |\Omega|.$$

Note that the function G is convex and therefore we have

$$G(t) \geq a_g t \text{ for all } t \geq 0.$$

Then, in view of (4.9) we are led to

$$J(u) \geq \left(\frac{a_g}{p} - c_1\lambda_1^{-1}\right) \|\nabla u\|_{L^p(\Omega)}^p - \frac{1}{q}|\Omega|^{\frac{p-q}{p}} \|\nabla u\|_{L^p(\Omega)}^q - c_1|\Omega|$$

for all $u \in W_0^{1,p}(\Omega)$. Since $q < p$ and $c_1 < \frac{a_g\lambda_1}{p}$, we obtain that J is bounded from below and coercive.

All the hypotheses required to apply Theorem 4.2 to the functional J in (4.9) are verified. As a consequence, the existence of a generalized variational solution to problem (4.1) with A given in (4.11) and $h = 0$ is established, whence the stated result for the original problem (4.7). $\qquad\qquad\square$

Remark 4.3 Corollary 4.1 does not apply to problem (4.7), but it can be applied for example when the driving operator

$$- \operatorname{div}\left(g(p^{-1}|\nabla u|^p)|\nabla u|^{p-2}\nabla u\right) + \operatorname{div}\left(|\nabla u|^{q-2}\nabla u\right)$$

is replaced by the negative (p, q)-Laplacian $-\Delta_p u - \Delta_q u$.

Chapter 5
Generalized Solutions for Inclusions

Abstract In this chapter, we continue our investigation of variational problems. However, unlike previously, we do not impose the differentiability assumption on the Euler action functional. Instead, we assume that the functional is locally Lipschitz continuous. As a result, we do not deal with a standard equation but rather with a nonlinear inclusion. For such an inclusion, we introduce a concept of generalized solution and obtain the relevant existence result. We provide applications to problems driven by the perturbed (p, q)-competing Laplacian, which we introduce here, and also to competing problems in variable Sobolev spaces.

Keywords Clarke derivative · Galerkin basis · Generalized competing (p, q)-Laplacian · Hemivariational inclusion · Non-smooth generalized solution · $p(x)$-Laplacian

5.1 Abstract Existence Results for Inclusions

In this chapter, for the theoretical setup we follow [34, 55], while we use other sources, mentioned later on, for examples which illustrate the theory.

The absence of continuous differentiability in the Euler action functional means that we cannot directly apply the abstract tools from the previous section. Therefore, we introduce abstract counterparts to the scheme outlined above. These counterparts are derived to handle the challenges posed by the lack of differentiability, allowing us to extend our analysis to a broader class of variational problems. Our abstract setting is as follows:

H(E)   *E is a separable, reflexive real Banach space densely and compactly embedded into another reflexive Banach space Y;*

H(A$_1$)   *A : E $\longrightarrow$ E*is a continuous, bounded and potential operator with the C^1 potential $\mathcal{A} : E \longrightarrow \mathbb{R}$, so $\mathcal{A}' = A$;*

H(A$_2$)   *$\Psi : Y \longrightarrow \mathbb{R}$ is a locally Lipschitz function such that the set*

© The Author(s), under exclusive license to Springer Nature Switzerland AG 2026

M. Galewski and D. Motreanu, *Competing Operators and Their Applications to Boundary Value Problems*, SpringerBriefs in Mathematics,

https://doi.org/10.1007/978-3-032-15445-3_5

$$\partial \Psi(B) = \bigcup_{v \in B} \partial \Psi(v)$$

is bounded for every bounded set $B \subset Y$.

H(A$_3$)   *The functional $J : E \to \mathbb{R}$ given by*

$$J(u) = \mathcal{A}(u) - \Psi(u) \text{ for all } u \in E$$

is coercive and bounded from below over E.

We recall that $\Psi : E \longrightarrow \mathbb{R}$ is a locally Lipschitz functional if every point $v \in E$ has an open neighborhood U_v and a constant $C_v > 0$ such that

$$|\Psi(w) - \Psi(y)| \le C_v \|w - y\| \quad \text{for all } w, y \in U_v.$$

The generalized directional derivative Ψ° of Ψ at $v \in E$ in the direction $w \in E$ is defined by

$$\Psi^\circ(v; w) = \limsup_{\substack{y \to v \\ t \downarrow 0}} \frac{\Psi(y + tw) - \Psi(y)}{t}$$

and the generalized gradient of Ψ at $v \in E$ is the set

$$\partial \Psi(v) := \left\{ \eta \in E^* : \Psi^\circ(v; w) \ge \langle \eta, w \rangle \text{ for every } w \in E \right\}.$$

For more details and an extended treatment on these notions we refer to [11–13, 47]. We examine the solvability of the following inclusion problem: find $u \in E$ such that

$$A(u) \in \partial \Psi(u). \tag{5.1}$$

This implies the hemivariational inequality

$$\langle Au, v \rangle \le \Psi^\circ(u; v) \quad \text{for all } v \in E. \tag{5.2}$$

The next definition describes the notion of generalized variational solution to problem (5.1) given so that it fits the nonlinear inclusion.

Definition 5.1 (*Abstract generalized solution*) Assume (H(E)), (H(A)$_1$) and (H(A)$_2$). An element $u \in E$ is said to be a generalized variational solution to problem (5.1) if there exists a sequence of subspaces $\{E_n\}_{n \ge 1} \subset E$ and a sequence of elements $\{u_n\}_{n \ge 1}$ for which

$$u_n \in E_n \text{ and } J(u_n) = \inf_{v \in E_n} J(v) \text{ for } n \in \mathbb{N}$$

and such that

(a) $u_n \rightharpoonup u$ in E as $n \to \infty$;
(b) $A(u_n) - z_n \rightharpoonup 0$ in E^*, with $z_n \in \partial \Psi(u_n)$;
(c) $\lim_{n \to \infty} \langle A(u_n), u_n - u \rangle = 0$.

The result below provides the existence of solutions to problem (5.1) in the described generalized sense.

Theorem 5.1 (Existence of generalized solutions for inclusion) *Assume (H(E)) and (H(A$_1$))-(H(A$_3$)). Then there exists at least one generalized variational solution $u_0 \in E$ to (5.1). If moreover the operator A satisfies property (S), then u_0 solves (5.1), in particular (5.2), in the usual weak sense.*

Proof Let a sequence of finite dimensional subspaces $\{E_n\}_{n \geq 1} \subset E$ form a Galerkin basis of E. We note that $\mathcal{A}$ is locally Lipschitz as it is C^1, thus the functional $J : E \to \mathbb{R}$ is locally Lipschitz. Accordingly, the functional J restricted to E_n is locally Lipschitz and coercive. Consequently, this restriction possesses at least one minimizer $u_n \in E_n$. Therefore for each n there exist $u_n \in E_n$ and $z_n \in E^*$ with $z_n \in \partial \Psi(u_n)$ such that

$$J(u_n) = \inf_{v \in E_n} J(v) \tag{5.3}$$

and

$$\langle A(u_n) - z_n, v \rangle = 0 \quad \text{for all } v \in E_n. \tag{5.4}$$

Moreover, since the space E is reflexive and the functional J is coercive, the sequence $\{u_n\}_{n \geq 1}$ is bounded in E and thus weakly convergent to some u_0 up to a subsequence. Condition (a) from Definition 5.1 is achieved. Through the density of $\cup_{n \geq 1} E_n$ in E we obtain from (5.4) that $A(u_n) - z_n \rightharpoonup 0$ in E^* which provides condition (b). By virtue of hypothesis (H(A$_2$)), the sequence $\{z_n\}_{n \geq 1}$ is bounded in E^*. The compact embedding of V into Y implies that $u_n \to u_0$ in Y, so we have

$$\lim_{n \to \infty} \langle z_n, u_n - u_0 \rangle = \lim_{n \to \infty} \langle z_n, u_n - u_0 \rangle_{Y^*, Y} = 0.$$

Then from (5.4) it follows that condition (c) holds true.

Now upon the assumption that A satisfies property (S), we obtain from (a) and (c) that $u_n \to u$ in E. Due to the fact that the sequence $\{z_n\}_{n \geq 1}$ is bounded in E^*, we may suppose that $z_n \rightharpoonup z_0$ for some $z_0 \in E^*$. Proposition 2.1.5 (b) in [11] allows us to assert that $z_0 \in \partial \Psi(u_0)$. Therefore by the continuity of A we see that

$$A(u_n) - z_n \rightharpoonup A(u_0) - z_0 = 0 \text{ in } E^*. \tag{5.5}$$

This concludes the proof. $\square$

Now we present a Weierstrass-Tonelli type result for locally Lipschitz functionals which are coercive and not necessarily sequentially weakly lower semicontinuous.

Theorem 5.2 (Weierstrass-Tonelli for locally Lipschitz functionals) *Assume (H(E)), (H(A)$_1$)-(H(A)$_3$). Then there is at least one generalized variational solution $u_0 \in E$ to (5.1). Moreover, there exist sequences $\{u_n\}_{n\geq 1}$ and $\{\xi_n\}_{n\geq 1}$ such that $u_n \in E_n$ for $n \in \mathbb{N}$ (where E_n is defined in Remark 2.11), $\xi_n \in \partial J(u_n)$,*

$$\inf_{u \in E_n} J(u) = J(u_n)$$

and

$$u_n \rightharpoonup u_0, \; J(u_n) \to \inf_{w \in E} J(w) \text{ and } \xi_n \rightharpoonup 0.$$

If moreover the operator A satisfies property (S), then u_0 solves (5.1) in the usual weak sense and the convergence of $\{u_n\}_{n\geq 1}$ is strong.

Proof Let us prove that the sequence $\{u_n\}_{n\geq 1} \subset E$ constructed in Theorem 5.1 is a minimizing sequence for the functional J, that is,

$$\lim_{n \to \infty} J(u_n) = \inf_{w \in E} J(w).$$

Indeed, from (5.3) and (H(A)$_3$) we see that the sequence $\{J(u_n)\}_{n\geq 1}$ is nonincreasing and bounded, so there exists the limit

$$l := \lim_{n \to \infty} J(u_n).$$

Arguing by contradiction, we suppose that

$$l > \inf_{w \in E} J(w).$$

Hence there is some $\hat{w} \in E$ such that $J(\hat{w}) < l$. The continuity of the functional J guarantees that there exists a neighborhood U of $\hat{w}$ in E such that

$$J(w) < l, \quad \forall w \in U$$

Since

$$U \cap \left(\bigcup_{n=1}^{\infty} E_n \right) \neq \emptyset,$$

we have that $\tilde{w} \in U \cap E_m$ for some m. This results in the contradiction

$$\min_{v \in E_m} J(v) \leq J(\tilde{w}) < l \leq \min_{v \in E_m} J(v),$$

thus the assertion is proven.

The fact that $\xi_n \to 0$, where $\xi_n \in \partial J(u_n)$, follows from (5.5), which completes the proof. $\square$

5.2 Applications to Boundary Value Problems

We proceed by applying the abstract results to a differential inclusion problem. Let a bounded domain $\Omega \subset \mathbb{R}^N$ with Lipschitz boundary $\partial\Omega$ and let $1 < q < p < +\infty$. We assume that

H(φ1) $\varphi_p, \varphi_q : \Omega \times \mathbb{R}_+ \to \mathbb{R}_+$ *are Carathéodory functions for which there are constants $M_p, M_q > 0$ such that*

$$\left| \varphi_p (x, u) \right| \leq M_p, \ \left| \varphi_q (x, u) \right| \leq M_q \ \textit{for a.e. } x \in \Omega \text{ and all } u \in \mathbb{R}_+$$

and which are homogenous with respect to u for a.e. $x \in \Omega$;

H(φ2)   *there exist constants $\gamma_p, \gamma_q > 0$ such that*

$$\varphi_p (x, u)\, u - \varphi_p (x, v)\, v \geq \gamma_p (u - v),$$

$$\varphi_q (x, u)\, u - \varphi_q (x, v)\, v \geq \gamma_q (u - v)$$

for all $u \geq v \geq 0$ and a.e. $x \in \Omega$.

Accordingly, we consider the operators

$$A_p, A_q : W_0^{1,p} (\Omega) \to W^{-1,p'} (\Omega)$$

given by

$$\langle A_p (u), v \rangle = \int_\Omega \varphi_p \left(x, |\nabla u (x)|^{p-1} \right) |\nabla u (x)|^{p-2}\, \nabla u (x)\, \nabla v (x)\, \mathrm{d}x, \qquad (5.6)$$

$$\langle A_q (u), v \rangle = \int_\Omega \varphi_q \left(x, |\nabla u (x)|^{q-1} \right) |\nabla u (x)|^{q-2}\, \nabla u (x)\, \nabla v (x)\, \mathrm{d}x. \qquad (5.7)$$

Lemma 5.1 *Assume that conditions $(H(\varphi 1))$-$(H(\varphi 2))$ are satisfied. Then the operator $A_p : W_0^{1,p} (\Omega) \to W^{-1,p'} (\Omega)$ defined in (5.6) is continuous, bounded, strictly monotone, coercive and satisfies condition (S). Moreover, it is potential with the potential $\mathcal{A}_p : W_0^{1,p} (\Omega) \to \mathbb{R}$ defined by*

$$\mathcal{A}_p (u) = \int_\Omega \int_0^{|\nabla u(x)|} \varphi_p \left(s^{p-1} \right) s^{p-1}\, ds\, dx \ \textit{for all } u \in W_0^{1,p} (\Omega). \qquad (5.8)$$

It also holds that A_p is d-monotone with respect to the function

$$\rho\left(t\right) = \gamma_p t^{p-1}$$

and A_p *is invertible with a continuous inverse. The operator* $A_q : W_0^{1,p}\left(\Omega\right) \to W^{-1,p'}\left(\Omega\right)$ *in* (5.7) *is continuous, bounded, and strictly monotone. It is also potential with the potential* $\mathcal{A}_q : W_0^{1,p}\left(\Omega\right) \to \mathbb{R}$ *defined by*

$$\mathcal{A}_q\left(u\right) = \int_{\Omega} \int_0^{|\nabla u(x)|} \varphi_q\left(s^{q-1}\right) s^{q-1} ds\, dx \ \text{for } u \in W_0^{1,p}\left(\Omega\right). \tag{5.9}$$

Proof The assertions about the operator A_p follow from Theorem 2.8 and from Theorem 2.9. For the operator A_q in (5.7) we cannot directly argue as above since $W_0^{1,p}\left(\Omega\right)$ is not the full domain of A_q. The strict monotonicity and the formula for the potential in (5.9) follow by a direct calculation, whereas the continuity is now due to the fact that $W_0^{1,p}\left(\Omega\right)$ embeds continuously into $W_0^{1,q}\left(\Omega\right)$. $\square$

Corollary 5.1 *Under assumptions* $(H(\varphi 1))$-$(H(\varphi 2))$, *we have the following estimates*

$$\frac{M_p}{p}\left\|u\right\|_{W_0^{1,p}(\Omega)}^p \geq \mathcal{A}_p\left(u\right) \geq \frac{\gamma_p}{p}\left\|u\right\|_{W_0^{1,p}(\Omega)}^p \tag{5.10}$$

and

$$\gamma_p\left\|u\right\|_{W_0^{1,p}(\Omega)}^p \leq \left\langle A_p\left(u\right), u\right\rangle \leq M_p\left\|u\right\|_{W_0^{1,p}(\Omega)}^p. \tag{5.11}$$

for all $u \in W_0^{1,p}\left(\Omega\right)$. *Similar estimates hold for* $\mathcal{A}_q$.

Proof Estimates (5.10) and (5.11) are direct consequences of (5.8) and (5.6), respectively. $\square$

Now we introduce a new operator that extends the competing (p, q)-Laplacian.

Definition 5.2 (*Perturbed competing* (p, q)-*Laplacian*) Assuming $(H(\varphi 1))$-$(H(\varphi 2))$, we call the operator $A_p - A_q : W_0^{1,p}\left(\Omega\right) \to W^{-1,p'}\left(\Omega\right)$ the perturbed competing (p, q)-Laplacian.

Remark 5.1 Such an operator as introduced in Definition 5.2 can be viewed as a type of Leray-Lions operator due to the presence of the term φ which satisfies conditions $(H(\varphi 1))$-$(H(\varphi 2))$ in its definition.

In the case

$$\varphi_p = \varphi_q = 1,$$

the perturbed competing (p, q)-Laplacian in Definition 5.2 reduces to the competing (p, q)-Laplacian $-\Delta_p + \Delta_q$.

Lemma 5.2 (Properties of perturbed competing (p, q)-Laplacian) *Under assumptions* $(H(\varphi 1))$-$(H(\varphi 2))$, *the operator* $A_p - A_q$ *is continuous, bounded and coercive.*

Proof The continuity and boundedness of the operator $A_p - A_q$ follow directly from Lemma 5.1. Concerning the coercivity, due to (5.11) and the inequality

$$\|u\|_{W_0^{1,q}(\Omega)} \le |\Omega|^{1/(p-q)} \|u\|_{W_0^{1,p}(\Omega)},$$

we see that for all $u \in W_0^{1,p}(\Omega)$ it holds

$$\langle A_p(u) - A_q(u), u \rangle \ge \gamma_p \|u\|_{W^{1,p}(\Omega)}^p - M_q |\Omega|^{1/(p-q)} \|u\|_{W^{1,p}(\Omega)}^q. \qquad (5.12)$$

Relation (5.12) implies the coercivity of $A_p - A_q$ because $p > q$. $\square$

We state the differential inclusion problem with Dirichlet boundary condition

$$\begin{cases} -\mathrm{div}(\varphi_p\left(x, |\nabla u|^{p-1}\right)|\nabla u|^{p-2}\nabla u) + \mathrm{div}(\varphi_q\left(x, |\nabla u|^{q-1}\right)|\nabla u|^{q-2}\nabla u) \in \partial F(u) \\ \quad \text{in } \Omega \\ u = 0 \quad \text{on } \partial\Omega. \end{cases}$$

$$(5.13)$$

The differential operator in (5.13) is the perturbed competing (p, q)-Laplacian

$$A_p - A_q : W_0^{1,p}(\Omega) \to W^{-1,p'}(\Omega),$$

where A_p and A_q are introduced in (5.6) and (5.7), respectively, for which conditions $(H(\varphi 1))$ and $(H(\varphi 2))$ are supposed to be satisfied.

In (5.13) we also have the generalized gradient ∂F of a locally Lipschitz function $F : \mathbb{R} \to \mathbb{R}$ satisfying the condition:

H(g₁) *There are positive constants c_0 and c_1 with $c_1 < \gamma_p \lambda_1$ such that*

$$|\zeta| \le c_0 + c_1|t|^{p-1} \text{ for all } t \in \mathbb{R} \text{ and } \zeta \in \partial F(t).$$

Under hypothesis (H(g)), the functional $\Psi : L^p(\Omega) \to \mathbb{R}$ given by

$$\Psi(v) = \int_\Omega F(v(x))\mathrm{d}x \text{ for all } v \in L^p(\Omega) \qquad (5.14)$$

is Lipschitz continuous on the bounded subsets of $L^p(\Omega)$. Hence it is well defined the generalized gradient $\partial\Psi : L^p(\Omega) \to 2^{L^{p'}(\Omega)}$ and condition $(H(A)_2)$ is satisfied. Through the compact embedding $W_0^{1,p}(\Omega) \hookrightarrow\hookrightarrow L^p(\Omega)$, it can be seen as a multivalued map

$$\partial\Phi : W_0^{1,p}(\Omega) \to 2^{W^{-1,p'}(\Omega)}.$$

With the function Ψ defined as in (5.14), we introduce the functional $J : W_0^{1,p}(\Omega) \to \mathbb{R}$ by

$$J(v) = \mathcal{A}_p(v) - \mathcal{A}_q(v) - \Psi(v) \text{ for all } v \in W_0^{1,p}(\Omega), \tag{5.15}$$

for the potentials $\mathcal{A}_p$ and $\mathcal{A}_q$ constructed in (5.8) and (5.9), respectively.

Lemma 5.3 *Assume conditions $(H(\varphi 1))$-$(H(\varphi 2))$ and $(H(g_1))$. Then the functional J given by (5.15) is locally Lipschitz with the generalized gradient*

$$\partial J(u) = A_p(u) - A_q(u) - \partial \Psi(u) \text{ for all } u \in W_0^{1,p}(\Omega) \tag{5.16}$$

and it is coercive and bounded from below on $W_0^{1,p}(\Omega)$.

Proof It is straightforward to check that the generalized gradient of J is given by formula (5.16). On the basis of (5.10) and $(H(g_1))$, we see for some constant $\widetilde{c} > 0$ that

$$J(u) \geq \frac{\gamma_p}{p} \|u\|_{W_0^{1,p}}^p - \frac{M_q}{q} |\Omega|^{\frac{p-q}{p}} \|u\|_{W_0^{1,p}}^q - \int_\Omega \left(c_0 |u(x)| + \frac{c_1}{p} |u(x)|^p \right) \mathrm{d}x$$

$$\geq \frac{1}{p} \left(\gamma_p - c_1 \lambda_1^{-1} \right) \|\nabla u\|_p^p - \frac{M_q}{q} |\Omega|^{\frac{p-q}{p}} \|\nabla u\|_p^q - \widetilde{c} \|\nabla u\|_p \text{ for all } u \in W_0^{1,p}(\Omega).$$

By hypothesis $c_1 < \gamma_p \lambda_1$, the functional J is coercive and bounded from below. $\square$

Our existence result concerning inclusion (5.13) reads as follows.

Theorem 5.3 (existence of generalized solutions for inclusions) *Assume conditions $(H(\varphi 1))$-$(H(\varphi 2))$ and $(H(g_1))$. Then problem (5.13) has at least one generalized variational solution.*

Proof Setting $E = W_0^{1,p}(\Omega)$, $Y = L^p(\Omega)$ and $A = A_p - A_q$, we note that condition $(H(A)_1)$ is satisfied. We have already shown above that condition $(H(A)_2)$ holds. Lemma 5.3 entails the validity of condition $(H(A)_2)$. We obtain the existence of at least one generalized variational solution to problem (5.13) due to Theorem 5.1, thus completing the proof. $\square$

For more developments, we refer to [34].

5.3 Applications for Problems in Variable Sobolev Spaces

Let $\Omega \subset \mathbb{R}^N (N \geq 2)$ be a bounded domain with a smooth boundary $\partial\Omega$. Let μ be a numerical parameter. Following [38] we study a Dirichlet μ-parametric differential inclusion of the form

$$-\Delta_{p(x)}u(x) + \mu\Delta_{q(x)}u(x) \in \partial F(u) \quad \text{in } \Omega, \quad u|_{\partial\Omega} = 0. \tag{5.17}$$

The variable exponents $p, q \in C(\bar{\Omega})$ are considered under the assumptions

$$1 < q^- = \inf_{x \in \bar{\Omega}} q(x) \leq q(x) \leq q^+ = \sup_{x \in \bar{\Omega}} q(x) <$$
$$p^- = \inf_{x \in \bar{\Omega}} p(x) \leq p(x) \leq p^+ = \sup_{x \in \bar{\Omega}} p(x) < +\infty.$$

The operator $-\Delta_{p(x)} + \mu \Delta_{q(x)}$, weighted by the parameter $\mu \in \mathbb{R}$, becomes competing for all $\mu < 0$, while for $\mu = 0$ it reduces to the $-\Delta_{p(x)}$-Laplacian and for $\mu > 0$ it becomes the variable (p, q)-Laplacian.

About the reaction we assume what follows where by C_{p^-} we mean the best Sobolev constant for the embedding of the Sobolev space $W_0^{1,p(x)}(\Omega)$ into $L^{p^-}(\Omega)$:

H(g_2) $F : \mathbb{R} \to \mathbb{R}$ is a locally Lipschitz function and there exist positive constants $c, \widehat{c} > 0$ with

$$\widehat{c} p^+ C_{p^-}^{p^-} < p^-, \tag{5.18}$$

such that

$$|z| \leq c + \widehat{c}|t|^{p^- - 1} \quad \text{for all } t \in \mathbb{R} \text{ and } z \in \partial F(t).$$

In the above we consider the generalized gradient (Clarke subdifferential) ∂F of $F : \mathbb{R} \to \mathbb{R}$, and we denote with F° its generalized directional derivative.

As in the previous section , we retrieve the following hemivariational inequality

$$\langle -\Delta_{p(x)} u, h \rangle + \mu \langle \Delta_{q(x)} u, h \rangle \leq \int_\Omega F^\circ(u(x); h(x)) dx \tag{5.19}$$

for all $h \in W_0^{1,p(x)}(\Omega)$, where the brackets $\langle \cdot, \cdot \rangle$ refer to the duality between the Banach spaces $W_0^{1,p(x)}(\Omega)$ and $\left(W_0^{1,p(x)}(\Omega) \right)^*$, i.e.

$$\langle -\Delta_{p(x)} u(x) + \mu \Delta_{q(x)} u(x), h(x) \rangle =$$
$$\int_\Omega |\nabla u(x)|^{p(x)-2} \nabla u(x) \cdot \nabla h(x) dx - \mu \int_\Omega |\nabla u(x)|^{q(x)-2} \nabla u(x) \cdot \nabla h(x) dx.$$

We say that a function $u \in W_0^{1,p(x)}(\Omega)$ is a weak solution to problem (5.17) if it solves the inequality (5.19).

On the other hand, writing explicitly we have the following notion:

Definition 5.3 (*Generalized solution*) We say that a function $u \in W_0^{1,p(x)}(\Omega)$ is a generalized solution to problem (5.17) if we can find a sequence $\{u_n\}_{n \geq 1} \subset W_0^{1,p(x)}(\Omega)$ satisfying

(a) $u_n \rightharpoonup u$ in $W_0^{1,p(x)}(\Omega)$, as $n \to +\infty$;

(b) $-\Delta_{p(x)} u_n + \mu \Delta_{q(x)} u_n - z_n \rightharpoonup 0$ in $\left(W_0^{1,p(x)}(\Omega) \right)^*$, as $n \to +\infty$, with $z_n \in \left(W_0^{1,p(x)}(\Omega) \right)^*$ and $z_n \in \partial F(u_n)$ a.e. in Ω;

(c) $\left\langle -\Delta_{p(x)}u_n + \mu\Delta_{q(x)}u_n, u_n - u \right\rangle \to 0$, as $n \to +\infty$.

Under hypothesis (H(g_2)), the functional $\Psi : L^{p(x)}(\Omega) \to \mathbb{R}$ given by

$$\Psi(v) = \int_\Omega F(v(x))dx \text{ for all } v \in L^{p(x)}(\Omega) \tag{5.20}$$

is Lipschitz continuous on the bounded subsets of $L^{p(x)}(\Omega)$. Moreover, it is well defined and its generalized gradient

$$\partial\Psi : L^{p(x)}(\Omega) \to 2^{L^{p'(x)}(\Omega)}$$

satisfies condition (H(A)$_2$). Due to the compact embedding of $W_0^{1,p(x)}(\Omega)$ into $L^{p(x)}(\Omega)$, see Proposition 1.3, we can consider it as a multivalued map

$$\partial\Phi : W_0^{1,p(x)}(\Omega) \to 2^{\left(W_0^{1,p(x)}(\Omega)\right)^*}.$$

As in the constant exponent case we introduce the functional $J : W_0^{1,p(x)}(\Omega) \to \mathbb{R}$ by

$$J(v) = \int_\Omega \frac{|\nabla u\,(x)\,|^{p(x)}}{p\,(x)}dx - \mu \int_\Omega \frac{|\nabla u\,(x)\,|^{q(x)}}{q\,(x)}dx - \int_\Omega F(v(x))dx \tag{5.21}$$

for all $v \in W_0^{1,p}(\Omega)$.

Lemma 5.4 *Let $\mu \in \mathbb{R}$ be fixed. Assume condition (H(g_2)). Then the functional J given by (5.21) is locally Lipschitz with the generalized gradient given by*

$$\partial J(u) = -\Delta_{p(x)}u + \mu\Delta_{q(x)}u - \partial\Psi(u) \tag{5.22}$$

for all $u \in W_0^{1,p(x)}(\Omega)$. Moreover, J is coercive and bounded from below on $W_0^{1,p(x)}(\Omega)$.

Proof It is straightforward to check, using (5.20), that the generalized gradient of J is given by formula (5.22). We need to show the coercivity of E. Using the continuous embedding $W_0^{1,p(x)}(\Omega) \hookrightarrow W_0^{1,q^-}(\Omega)$, and Remark 1.3 we have

$$\int_\Omega |\nabla u|^{q(x)}dx \leq \|\nabla u\|_{q(x)}^{q^-} + 1$$
$$\leq C\|\nabla u\|_{p(x)}^{q^-} + 1,$$

where the constant $C > 0$ refers to the continuous embedding of $W_0^{1,p(x)}(\Omega)$ in $W_0^{1,q^-}(\Omega)$. Using assumption (H(g_2)) and the mean-value theorem, we obtain

$$|F(t)| \leq |F(0)| + c|t| + \frac{\widehat{c}}{p^-}|t|^{p^-} \quad \text{for all } t \in \mathbb{R}.$$

The above and Remark 1.3 lead to the following estimate:

$$
\begin{aligned}
J(u) &\geq \tfrac{1}{p^+}\int_\Omega |\nabla u|^{p(x)}dx - \tfrac{|\mu|}{q^-}\int_\Omega |\nabla u|^{q(x)}dx - \int_\Omega \left(c|u| + \tfrac{\widehat{c}}{p^-}|u|^{p^-}\right)dx - |F(0)||\Omega| \\
&\geq \tfrac{1}{p^+}\left(\|\nabla u\|_{p(x)}^{p^-} - 1\right) - \tfrac{|\mu|}{q^-}\left(C\|\nabla u\|_{p(x)}^{q^+} + 1\right) \\
&\quad - cC_1\|\nabla u\|_{p(x)} - \tfrac{\widehat{c}}{p^-}C_{p^-}^{p^-}\|\nabla u\|_{p(x)}^{p^-} - |F(0)||\Omega| \\
&\geq \tfrac{1}{p^+}\left(1 - \tfrac{\widehat{c}p^+}{p^-}C_{p^-}^{p^-}\right)\|\nabla u\|_{p(x)}^{p^-} - \tfrac{|\mu|}{q^-}C\|\nabla u\|_{p(x)}^{q^+} - cC_1\|\nabla u\|_{p(x)} - \tfrac{1}{p^+} - \tfrac{|\mu|}{q^-} - |F(0)||\Omega|,
\end{aligned}
$$

where C_1 is a positive constant, $|\Omega|$ is the Lebesgue measure of Ω. Since $\widehat{c}p^+ C_{p^-}^{p^-} < p^-$ and since $p^- > q^+$, we conclude that the functional J is coercive. The same argument shows that it is bounded from below. $\qquad\square$

Our results on inclusion (5.17) read as follows:

Theorem 5.4 *Let $\mu \geq 0$. Assume condition $(H(g_2))$. Then problem (5.17) has at least one generalized variational solution which is not trivial provided that $0 \notin \partial F(u)$ on a set of positive measure for any $u \in L^{p(x)}(\Omega)$.*

Proof Setting $E = W_0^{1,p(x)}(\Omega)$, $Y = L^{p(x)}(\Omega)$ and

$$A(u) = \int_\Omega \frac{1}{p(x)}|\nabla u(x)|^{p(x)}dx - \mu\int_\Omega \frac{1}{q(x)}|\nabla u(x)|^{q(x)}dx,$$

we note that condition $(H(A)_1)$ is satisfied. We have already shown above that condition $(H(A)_2)$ holds. Lemma 5.4 entails the validity of condition $(H(A)_2)$. We obtain the existence of at least one generalized variational solution to problem (5.17) due to Theorem 5.1, thus completing the proof. $\qquad\square$

Proposition 5.1 *Let $\mu < 0$. Assume condition $(H(g))$. Then problem (5.17) has at least one weak solution.*

Proof Lemma 5.4 tells us that the functional J is coercive. Moreover, it is sequentially weakly lower semicontinuous and therefore by the Weierstrass-Tonelli Theorem it has at least one minimizer which due to the non-smooth Fermat rule satisfies (5.19) in the weak sense.

For the proof we can also apply the second part of Theorem 5.1, since the operator

$$-\Delta_{p(x)} + \mu\Delta_{q(x)}$$

is now of type $(S)_+$. $\qquad\square$

Remark 5.2 We can alter assumption (H(g_2)) through a more general growth condition of the following form: there exist a function $\alpha \in C(\bar{\Omega})$ and positive constants $c, \widehat{c} > 0$ with $\widehat{c} p^+ C_{p^-}^{p^-} < p(x)$, such that

$$|z| \leq c + \widehat{c}|t|^{\alpha(x)} \quad \text{for all } t \in \mathbb{R} \text{ and } z \in \partial F(t)$$

This setting requires the additional inequalities

$$1 \leq \alpha(x) \leq \alpha^+ \leq p^- \quad \text{for all } x \in \bar{\Omega}$$

to conclude that the functional is coercive in Lemma 5.4. However, if $\alpha^+ < p^-$ then there is no need to assume inequality (5.18). The proofs remain the same in essence, up to some technical details.

For more developments on systems of variable inclusions, we refer to [19]. We summarize these results below:

Remark 5.3 Consider the following problem of differential inclusions with Dirichlet boundary condition

$$\begin{cases} \left(-\Delta_{p_1(\cdot)}u_1 + \mu_1\Delta_{q_1(\cdot)}u_1, -\Delta_{p_2(\cdot)}u_2 + \mu_2\Delta_{q_2(\cdot)}u_2\right) \in \partial F(u_1, u_2) & \text{in } \Omega \\ u_1 = u_2 = 0 & \text{on } \partial\Omega \end{cases}$$
$$(5.23)$$

where $\mu_1, \mu_2 \in \mathbb{R}$ are parameters and $-\Delta_{p_i(\cdot)}$ and $\Delta_{q_i(\cdot)}$, for $i = 1, 2$, denote the negative $p_i(\cdot)$-Laplace operator and the positive $q_i(\cdot)$-Laplace operator, respectively.
Here $p_i, q_i \in C(\bar{\Omega})$ are such that

$$1 < q_i^- \leq q_i(x) \leq q_i^+ < p_i^- \leq p_i(x) \leq p_i^+ < +\infty$$

for all $x \in \bar{\Omega}$ and $i = 1, 2$. In the right-hand side of problem (5.23), we find the generalized gradient ∂F of a locally Lipschitz function $F : \mathbb{R}^2 \to \mathbb{R}$ and we note that pointwise $\partial F(u_1, u_2)$ is a subset of $\mathbb{R}^2$. Therefore, (5.23) is a system of two hemivariational inclusions. Corresponding to (5.23) we have the following hemivariational inequality

$$\begin{aligned} &\left\langle-\Delta_{p_1(\cdot)}u_1, h_1\right\rangle + \mu_1\left\langle\Delta_{q_1(\cdot)}u_1, h_1\right\rangle + \left\langle-\Delta_{p_2(\cdot)}u_2, h_2\right\rangle + \mu_2\left\langle\Delta_{q_2(\cdot)}u_2, h_2\right\rangle \leq \\ &\int_\Omega F^\circ(u_1, u_2; h_1, h_2)\,\mathrm{d}x \end{aligned}$$
$$(5.24)$$

for all $(h_1, h_2) \in W_0^{1, p_1(\cdot)}(\Omega) \times W_0^{1, p_2(\cdot)}(\Omega)$, with F° being the generalized directional derivative of F. The assumption on F is as follows: there exist positive constants $a_0, a_1, a_2, b_0, b_1, b_2, \alpha_1, \alpha_2$ with

$$1 < \alpha_1 < p_1^-, \quad 1 < \alpha_2 < p_2^-, \quad a_1 p_1^+ A_1 < p_1^-$$

and

$$b_2 p_2^+ B_1 < p_2^-,$$

such that

$$|z_1| \le a_0 + a_1 |t|^{p_1^- - 1} + a_2 |s|^{\frac{p_2^-}{\alpha_1}}$$

and

$$|z_2| \le b_0 + b_1 |t|^{\frac{p_1^-}{\alpha_2}} + b_2 |s|^{p_2^- - 1}$$

for all $t, s \in \mathbb{R}$ and $(z_1, z_2) \in \partial F(t, s)$, where A_1 and B_1 are the best constants such that

$$\int_\Omega |u_1|^{p_1^-} \, dx \le A_1 \, \|\nabla u_1\|_{p_1(\cdot)}^{p_1^-} \quad \text{for all } u_1 \in W_0^{1, p_1(\cdot)}(\Omega),$$

$$\int_\Omega |u_2|^{p_2^-} \, dx \le B_1 \, \|\nabla u_2\|_{p_2(\cdot)}^{p_2^-} \quad \text{for all } u_2 \in W_0^{1, p_2(\cdot)}(\Omega).$$

Under these conditions if $\mu_1 \le 0$ and $\mu_2 \le 0$, then any generalized solution to problem (5.23) is in addition a weak solution, i.e. satisfies (5.24). Moreover, for any $\mu_1, \mu_2 \in \mathbb{R}$ problem (5.23) admits at least a generalized solution (u_1, u_2) in $W_0^{1, p_1(\cdot)}(\Omega) \times W_0^{1, p_2(\cdot)}(\Omega)$.

We conclude with some standard examples of functions satisfying the conditions given in Remark 5.3.

Example 5.1 Let a locally Lipschitz function $f_1 : \mathbb{R} \to \mathbb{R}$ be given by

$$f_1(t) = \begin{cases} t^2 \cos \frac{1}{t} & \text{if } t \ne 0 \\ 0 & \text{if } t = 0 \end{cases}$$

with the Clarke subdifferential

$$\partial f_1(t) = \begin{cases} 2t \cos \frac{1}{t} + \sin \frac{1}{t} & \text{if } t \ne 0 \\ [-1, 1] & \text{if } t = 0 \end{cases}$$

Consider another locally Lipschitz function $f_2 : \mathbb{R} \to \mathbb{R}$ defined by

$$f_2(t) = |t| \quad \text{for all } t \in \mathbb{R}.$$

with the Clarke subdifferential

$$\partial f_2 = \begin{cases} -1 & \text{if } t < 0 \\ [-1, 1] & \text{if } t = 0 \\ 1 & \text{if } t > 0 \end{cases}$$

Using f_1 and f_2, we introduce the function $F : \mathbb{R}^2 \to \mathbb{R}$ defined by

$$F(t, s) = f_1(t) + t f_2(s) + s^2$$

for all $t, s \in \mathbb{R}$. Then F is a locally Lipschitz function and the Clarke subdifferential of F is given by

$$\partial F(t, s) = (\partial f_1(t) + f_2(s), t \partial f_2(s) + 2s)$$

for all $t, s \in \mathbb{R}$. Taking this into account, for any $(\zeta_1, \zeta_2) \in \partial F(t, s)$ we have that the following inequalities

$$|\zeta_1| \le 1 + 2|t| + |s| \quad \text{and} \quad |\zeta_2| \le |t| + 2|s|$$

are verified for all $(t, s) \in \mathbb{R}^2$. Then for $\alpha_i = 2$ and $p_i^- > 2$ for $i = 1, 2$ the growth assumptions are satisfied.

Bibliography

1. Adams, R.A.: Sobolev Spaces. Academic Press, New York (1975)
2. Andrzejczak, G., Galewski, M., Motreanu, D.: Existence theorems for parameter dependent weakly continuous operators with applications. Results Math. 79(4), 20 (Paper No. 160) (2024)
3. de Araujo, A.L., Medeiros, A.H., Motreanu, D.: Solutions for a nonlinear equation with competing operators and supercritical exponential growth. Complex Var. Elliptic Equ. **70**(5), 820–833 (2025)
4. Brasco, L., Lindgren, E., Parini, E.: The fractional Cheeger problem. Interfaces Free Bound. **16**, 419–458 (2014)
5. Brasco, L., Parini, E.: The second eigenvalue of the fractional p-Laplacian. Adv. Calc. Var. **9**, 323–355 (2016)
6. Borer, F., Carl, S., Winkert, P.: Elliptic p-Laplacian systems with nonlinear boundary condition. J. Math. Anal. Appl. 538(2), 28 (Paper No. 128421) (2024)
7. Chiappinelli, R., Edmunds, D.E.: Remarks on surjectivity of gradient operators. Mathematics **8**, 1538 (2020)
8. Brézis, H.: Functional Analysis. Sobolev Spaces and Partial Differential Equations. Springer, New York (2010)
9. Cen, J., Marano, S., Zeng, S.: Differential inclusion systems with double phase competing operators, convection, and mixed boundary conditions. Appl. Math. Lett. 167, 6 (Paper No. 109556) (2025)
10. Chabrowski, J.: Variational Methods for Potential Operator Equations. De Gruyter, Berlin (1997)
11. Clarke, F.H.: Optimization and Nonsmooth Analysis. Wiley, New York (1983)
12. Denkowski, Z., Migórski, S., Papageorgiou, N.S.: An Introduction to Nonlinear Analysis: Theory. Kluwer Academic/Plenum Publishers, New York (2003)
13. Denkowski, Z., Migórski, S., Papageorgiou, N.S.: An Introduction to Nonlinear Analysis: Applications. Kluwer Academic/Plenum Publishers, New York (2003)
14. Diening, L., Harjulehto, P., Hästö, P., Růžička, M.: Lebesgue and Sobolev Spaces with Variable Exponents. Lecture Notes in Mathematics, vol. 2017. Springer, Berlin (2011)
15. Diblik, J., Galewski, M., Kossowski, I., Motreanu, D.: On competing (p, q)-Laplacian Dirichlet problem with unbounded weight. Differ. Integral Equ. **38**(1–2), 23–42 (2025)
16. Dinca, G., Jebelean, P., Mawhin, J.: Variational and topological methods for Dirichlet problems with p-Laplacian. Port. Math. (N.S.) 58(3), 339–378 (2001)
17. Drábek, P., Kufner, A., Nicolosi, F.: Quasilinear Elliptic Equations with Degenerations and Singularities, De Gruyter Series in Nonlinear Analysis and Applications, vol. 5. Walter de Gruyter, Berlin (1997)

© The Editor(s) (if applicable) and The Author(s), under exclusive license to Springer Nature Switzerland AG 2026

M. Galewski and D. Motreanu, *Competing Operators and Their Applications to Boundary Value Problems*, SpringerBriefs in Mathematics,
https://doi.org/10.1007/978-3-032-15445-3

18. Drábek, P., Milota, J.: Methods of Nonlinear Analysis: Applications to Differential Equations, 2nd edn. Birkhäuser, Basel (2013)

19. Efendiev, R., Vetro, F.: Systems of differential inclusions with competing operators and variable exponents. Opuscula Math. **45**(5), 665–684 (2025). https://doi.org/10.7494/OpMath.2025.45.5.665

20. Edmunds, D.E., Evans, W.D.: Fractional Sobolev Spaces and Inequalities, Cambridge Tracts in Mathematics, vol. 230. Cambridge University Press, Cambridge (2023)

21. Ekeland, I.: Nonconvex minimization problems. Bull. Amer. Math. Soc. (N.S.) 1(3), 443–474 (1979)

22. Ekeland, I., Temam, R.: Convex Analysis and Variational Problems. North-Holland, Amsterdam (1976)

23. Figuereido, D.G.: Lectures on the Ekeland Variational Principle with Applications and Detours. Preliminary Lecture Notes. SISSA, Trieste (1988)

24. Figueiredo, G.M., Razani, A.: Degenerated and competing horizontal (p, q)-Laplacians with weights on the Heisenberg group. Numer. Funct. Anal. Optim. **44**, 179–201 (2023)

25. Figueiredo, G.M., Razani, A.: Degenerated and competing anisotropic (p, q)-Laplacians with weights. Appl. Anal. **102**, 4471–4488 (2023)

26. Figueiredo, G.M., Vetro, C.: The existence of solutions for the modified $(p(x), q(x))$-Kirchhoff equation, Electron. J. Qual. Theory Differ. Equ. 2022, 16 (Paper No. 39)

27. Franců, J.: Monotone operators: a survey directed to applications to differential equations. Aplikace Mat. **35**(4), 257–301 (1990)

28. Frassu, S., Iannizzotto, A.: Extremal constant sign solutions and nodal solutions for the fractional p-Laplacian. J. Math. Anal. Appl. 501 (Paper No. 124205) (2021)

29. Fučík, S., Kufner, A.: Nonlinear Differential Equations, Studies in Applied Mechanics, vol. 2. Elsevier, Amsterdam (1980)

30. Gajewski, H., Gröger, K., Zacharias, K.: Nichtlineare Operatorgleichungen und Operatordifferentialgleichungen. Akademie-Verlag, Berlin (1974)

31. Galewski, M.: Basic Monotonicity Methods with Some Applications. Compact Textbooks in Mathematics. Birkhäuser; SpringerNature, Cham (2021)

32. Galewski, M., Motreanu, D.: On variational approach to fourth-order problems with unbounded weight. Math. Methods Appl. Sci. **46**(18), 18579–18591 (2023)

33. Galewski, M., Motreanu, D.: On variational competing (p, q)-Laplacian Dirichlet problem with gradient depending weight. Appl. Math. Lett. 148C, 7 (Article ID 108881) (2024)

34. Galewski, M., Motreanu, D.: On a hemivariational inequality with non-monotone operator. Optimization **74**(14), 3687–3701 (2025). https://doi.org/10.1080/02331934.2024.2385651

35. Gambera, L., Marano, S.A., Motreanu, D.: Quasilinear Dirichlet systems with competing operators and convection. J. Math. Anal. Appl. **530**(2) 11 (Paper No. 127718) (2024)

36. Gambera, L., Marano, S., Motreanu, D.: Dirichlet problems with fractional competing operators and fractional convection. Fract. Calc. Appl. Anal. **27**(5), 2203–2218 (2024)

37. Gasinski, L., Winkert, P.: Existence and uniqueness results for double phase problems with convection term. J. Differ. Equ. **268**, 4183–4193 (2021)

38. Ghasemi, M., Vetro, C., Zhang, Z.: Dirichlet μ-parametric differential problem with multivalued reaction term. Mathematics **13**(8), 1295 (2023). https://doi.org/10.3390/math13081295

39. Haase, M.: Functional Analysis: An Elementary Introduction, Graduate Studies in Mathematics, vol. 156. AMS, Providence (2014)

40. Iannizzotto, A., Mosconi, S., Squassina, M.: Global Hölder regularity for the fractional p-Laplacian. Rev. Mat. Iberoam. **32**, 1353–1392 (2016)

41. Lindgren, E., Lindqvist, P.: Fractional eigenvalues. Calc. Var. Partial. Differ. Equ. **49**, 795–826 (2014)

42. Liu, Y., Liu, Z., Motreanu, D.: Differential inclusion problems with convolution and discontinuous nonlinearities. Evol. Equ. Control Theory **9**, 1057–1071 (2020)

43. Liu, Z., Livrea, R., Motreanu, D., Zeng, S.: Variational differential inclusions without ellipticity condition. Electron. J. Qual. Theory Differ. Equ. **43**, 1–17 (2020)

44. Mawhin, J.: Problèmes de Dirichlet variationnels non linéaires, Séminaire de Mathématiques Supérieures, vol. 104, Montreal (1987)
45. Matei, A., Sofonea, M.: Variational Inequalities with Applications: A Study of Antiplane Frictional Contact Problems, Advances in Mechanics and Mathematics, vol. 18. Springer, New York (2009)
46. Medeiros, A.H.S., Motreanu, D.: A problem involving competing and intrinsic operators. Sao Paulo J. Math. Sci. 18(1), 300–311 (2024)
47. Migorski, S., Sofonea, M.: Variational-Hemivariational Inequalities with Applications. Monographs and Research Notes in Mathematics. CRC Press, Boca Raton (2018)
48. Mironescu, P., Sickel, W.: A Sobolev non embedding. Atti Accad. Naz. Lincei Cl. Sci. Fis. Mat. Natur. Rend. Lincei (9) Mat. Appl. 26, 291–298 (2015)
49. Molica Bisci, G., Rădulescu, V.D., Servadei, R.: Variational Methods for Nonlocal Fractional Problems, Encyclopedia of Mathematics and its Applications, vol. 162. Cambridge University Press, Cambridge (2016)
50. Motreanu, D.: Nonlinear Differential Problems with Smooth and Nonsmooth Constraints. Academic Press, London (2018)
51. Motreanu, D.: Quasilinear Dirichlet problems with competing operators and convection. Open Math. 18, 1510–1517 (2020)
52. Motreanu, D., Motreanu, V.V.: Nonstandard Dirichlet problems with competing (p, q)-Laplacian, convection, and convolution. Stud. Univ. Babeş-Bolyai Math. 66(1), 95–103 (2021)
53. Motreanu, D.: Nonhomogeneous Dirichlet problems with unbounded coefficient in the principal part. Axioms 11(12), 739 (2022)
54. Motreanu, D.: Equations with s-fractional (p, q)-Laplacian and convolution. Minimax Theory Appl. 7, 159–172 (2022)
55. Motreanu, D.: Hemivariational inequalities with competing operators. Commun. Nonlinear Sci. Numer. Simul. 130, 7 (Paper No. 107741) (2024)
56. Motreanu, D., Nashed, M.Z.: Degenerated (p, q)-Laplacian with weights and related equations with convection. Numer. Funct. Anal. Optim. 42(15), 1757–1767 (2021)
57. Motreanu, D., Rădulescu, V.D.: Variational and Nonvariational Methods in Nonlinear Analysis and Boundary Value Problems. Nonconvex Optimization and Its Applications. Springer, New York (2003)
58. Motreanu, D., Razani, A.: Competing anisotropic and Finsler (p, q)-Laplacian problems. Bound. Value Probl. 2024, 10 (Paper No. 39)
59. Motreanu, D., Razani, A.: Optimizing solutions with competing anisotropic (p, q)-Laplacian in hemivariational inequalities. Constr. Math. Anal. 7(4), 150–159 (2024)
60. Motreanu, D., Winkert, P.: Existence and asymptotic properties for quasilinear elliptic equations with gradient dependence. Appl. Math. Lett. 95, 78–84 (2019)
61. Di Nezza, E., Palatucci, G., Valdinoci, E.: Hitchhiker's guide to the fractional Sobolev spaces. Bull. Sci. Math. 136, 521–573 (2012)
62. Papageorgiou, N.S., Rădulescu, V.D., Repovš, D.D.: Nonlinear Analysis-Theory and Methods. Springer Monographs in Mathematics. Springer, Cham (2019)
63. Phelps, R.R.: Convex Functions, Monotone Operators and Differentiability. Lecture Notes in Mathematics, vol. 1364, 2nd edn. Springer, Berlin (1993)
64. Rădulescu, V.D., Repovš, D.D.: Partial Differential Equations with Variable Exponents: Variational Methods and Qualitative Analysis. Monographs and Research Notes in Mathematics. CRC Press, Boca Raton (2015)
65. Rudin, W.: Principles of Mathematical Analysis, 2nd edn. McGraw-Hill, New York (1964)
66. Razani, A.: Nonstandard competing anisotropic (p, q)-Laplacians with convolution. Bound. Value Probl. 2022, 10 (Paper No. 87)
67. Razani, A., Sevim, E.S., Babaei, S.: Competing Finsler double phase equation. Taiwanese J. Math. 29(3), 575–589 (2025)
68. Razani, A.: A solution of a nonstandard Dirichlet Finsler (p, q)-Laplacian. Filomat 38(23), 8131–8139 (2024)

69. Razani, A.: Competing Kohn-Spencer Laplacian systems with convection in non-isotropic Folland-Stein space. Complex Var. Elliptic Equ. **70**(4), 667–680 (2025)
70. Eddine, N.C., Ragusa, M.A.: The compact embeddings and the concentration-compactness principles for anisotropic variable exponent Sobolev spaces and applications. J. Geom. Anal. 35 (Article No. 338) (2025)
71. Rudin, W.: Functional Analysis. McGraw-Hill Series in Higher Mathematics. McGraw-Hill, New York (1973)
72. Roubíček, T.: Nonlinear Partial Differential Equations with Applications. International Series of Numerical Mathematics, vol. 153. Birkhäuser, Basel (2013)
73. Showalter, R.E.: Monotone Operators in Banach Space and Nonlinear Partial Differential Equations. American Mathematical Society, Providence (1997)
74. Troutman, J.L.: Variational Calculus and Optimal Control: Optimization with Elementary Convexity. Undergraduate Texts in Mathematics. Springer, New York (1996)
75. Zeidler, E.: Nonlinear Functional Analysis and Its Applications II/B: Nonlinear Monotone Operators. Springer, New York (1990)

Index

M. Galewski and D. Motreanu, *Competing Operators and Their Applications to Boundary
Value Problems*, SpringerBriefs in Mathematics,
https://doi.org/10.1007/978-3-032-15445-3